Architectural Design with SketchUp

Architectural Design with SketchUp

Component-Based Modeling, Plugins, Rendering, and Scripting

Alexander C. Schreyer

WILEY

John Wiley & Sons, Inc.

Cover design: David Riedy

Cover illustrations: (front and back) Courtesy of Alexander C. Schreyer

This book is printed on acid-free paper. ∞

For general information on our other products and services, or technical support, please contact our Customer Care Department within the United States at 800-762-2974, outside the United States at 317-572-3993 or fax 317-572-4002.

Wiley publishes in a variety of print and electronic formats and by print-on-demand. Some material included with standard print versions of this book may not be included in e-books or in print-on-demand. If this book refers to media such as a CD or DVD that is not included in the version you purchased, you may download this material at http://booksupport.wiley.com. For more information about Wiley products, visit our Web site at www.wiley.com.

Library of Congress Cataloging-in-Publication Data:

Schreyer, Alexander (Alexander C.)

Architectural Design with SketchUp: Component-based modeling, plugins, rendering, and scripting / Alexander Schreyer.
   p. cm.
Includes index.
ISBN 978-1-118-12309-6 (pbk.), ISBN 978-1-118-37652-2 (ebk.); ISBN 978-1-118-37654-6 (ebk.); ISBN 978-1-118-38596-8 (ebk.); ISBN 978-1-118-38597-5 (ebk.); ISBN 978-1-118-38598-2 (ebk.); ISBN 978-1-118-56787-6 (ebk.); ISBN 978-1-118-56776-0 (ebk.)
1. Computer graphics. 2. SketchUp. 3. Three-dimensional display systems. 4. Engineering graphics. I. Title.
T385.S3347 2013
006.6'8—dc23
                  2012007297

Printed in the United States of America

10 9 8 7 6 5 4 3 2 1

For my father, Gerhard

Contents

Appendices

Acknowledgments

Having taught SketchUp to varied audiences of eager students, I should start my acknowledgments with exactly those students whose many questions and creative ideas have inspired me not only to look deeper into the software, but also to put this text down on paper. *Keep pushing the boundaries of the third dimension in your work!*

An amazing product can often be judged by the community that develops around it. SketchUp has always been a small, yet transformative piece of software, whose simplicity and power have enthralled users for many years. This has created a large user community, which in forums, blogs, at user meetings, and other venues has—often passionately—taken to using it to design whatever came to their creative minds and educating others in how to use it to realize their ideas. I would like to hereby acknowledge that community for its devotion, support and inventiveness, and am with this book paying forward any support they ever gave me.

Among the makers of SketchUp I would like to foremost thank SketchUp product manager John Bacus and product evangelist Aidan Chopra for their feedback whenever I had a request—*and of course for the great time I had in Boulder.*

This book would not have been possible without the support and feedback from acquisitions editor Paul Drougas at John Wiley & Sons. This being my first book endeavor, I am still in awe of the amount of work that the editorial team puts into a publication like this. In particular, I would like to acknowledge production editor Nancy Cintron's tireless suggestions of edits and revisions as well as copyeditor Ginny Carroll's and editorial assistant Mike New's help in this process. Judging by the editing initials in the manuscript, it passed through many more hands whose anonymous work I hereby gratefully acknowledge.

Finally—and most importantly—I would like to thank the love of my life, my wonderful wife Peggi, and our two girls, Sophia and Mackenzie, for their love, tremendous encouragement, and support, as well as their patience with me while I was preparing the manuscript. I couldn't have done it without them! They are, together with my mother and my brother, the source of all my strength and joy.

Chapter 1
Introduction

In my years of teaching SketchUp, as well as other Computer-Aided Design (CAD) and Building Information Modeling (BIM) software, I have seen very proficient users of this software. Students and professionals take easily to SketchUp, and, before long, some of them produce very detailed building models and professional-grade renderings. But I have also found that too many people don't go beyond the basics and believe that some of the advanced modeling (or good-quality photorealistic rendering) needs to be done using other software. Very often, they painstakingly pick up that other software only to find that it is too complex, which likewise leaves them unable to do what they wanted.

Sometimes even advanced users of SketchUp master one aspect of the software (photorealistic rendering, for example) but are completely unaware of the power that SketchUp holds in other areas—Dynamic Components and Ruby scripting are good examples. As you will find out in this book, SketchUp is a very powerful design and 3D modeling tool. Some of its core features—for example, its extendibility with plugins—make it flexible enough to be useful for a variety of disciplines. The large number of high-quality plugins that are available for SketchUp these days bears powerful witness to this.

This book attempts to help the basic to intermediate user make the leap from simply creating "something" in SketchUp to using it as a powerful design tool. While it contains some more involved topics (such as photorealistic rendering, Dynamic Components, and Ruby scripting), it provides a clear learning path that takes you through easy-to-follow examples to a thorough understanding of the core topics. **Figure 1.1** shows an example of how one could use Ruby scripting to create geometry, then render the scene using a photorealistic rendering software and finally "dress-up" the image as a watercolor painting.

About This Book

Each chapter in this book presents a different SketchUp use in sufficient detail to get you started and working quickly. Interspersed with the text are many step-by-step examples, tips, and in-depth articles. At the end of each chapter, you will also find a collection of activities that you can undertake to try out new skills that you just learned.

Figure 1.1: Watercolor of a rendering of script-generated panels

Chapter 2, which follows this introductory chapter, brings every reader up to speed. Its purpose as a "SketchUp Refresher" is to review some basic modeling techniques and teach good practices for modeling and software use. While many readers will already have some knowledge of SketchUp through introductory books or video tutorials, this chapter encompasses enough variety to be useful for everyone, independent of their skill level.

Chapter 3 uses SketchUp not only as a modeling tool but also as a tool to inform your designs. In this chapter, you will learn more ways to employ SketchUp as an aid in your design process. Examples of this are creating component-based models, using Dynamic Components, and geo-based modeling. One section also looks at how SketchUp can fit into a BIM-based architectural design process.

Chapter 4 leads you into the wide field of SketchUp plugins and their uses. After an introductory section on finding and installing plugins, many individual plugins are discussed. Those small software add-ons to SketchUp provide tools for general modeling, such as drawing splines and lofting curves; tools for architectural design, such as stair making and wood framing; and tools for digital fabrication that will help you prepare your model for 2D and 3D digital printing and assembly. (See **Figure 1.2** for an example of a fabricated SketchUp model). Furthermore, there are plugins for data integration that can import or export data such as LIDAR laser-scan points, plugins for animation and presentation that add object animation or serve as helpers for creating animations and walk-throughs using SketchUp, and, finally, plugins for analysis, which provide analytical tools—mainly from the fields of building energy analysis and green building.

Chapter 5 introduces photorealistic rendering and covers all aspects of rendering in detail (see **Figure 1.3**). This chapter was written to be as independent of your actual choice of rendering software as possible, thus providing a useful resource no matter which software you download or buy. As part of this chapter, you will learn about modeling for rendering, lighting, sky environment, materials, and objects, as well as how to edit and modify renderings using image-editing software.

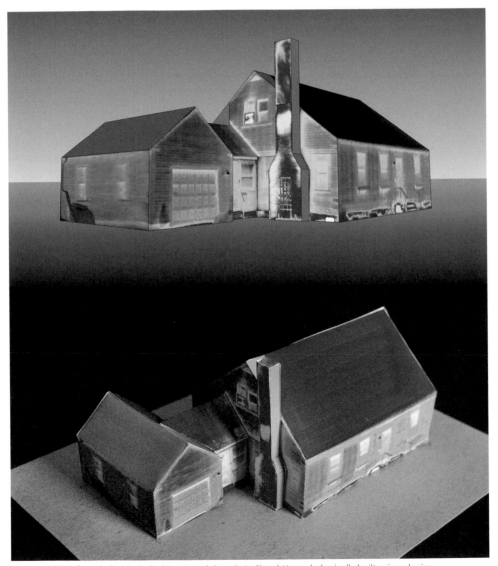

Figure 1.2: Infrared photography house model made in SketchUp and physically built using plugins

The final chapter in this book (Chapter 6) introduces you to the exciting field of computational geometry in SketchUp. This chapter presents Ruby Script examples that create undulating brick walls, solar-responsive facades, attractor-based colorful building designs, and other fun ways to create geometry in SketchUp without excessive use of the mouse (see **Figure 1.4**). Most of the script examples accomplish their tasks in just a few lines of code, and all are a good introduction to both the Ruby scripting language and the general field of computational geometry.

Figure 1.3: Glasses and liquid, rendered in SketchUp

Figure 1.4: A grassy hill made in SketchUp

As you will see in the chapters that follow, this book is intended to serve as a textbook as well as a desk reference. It was written to convey the presented material in a thorough yet easy-to-follow manner. It also covers common tasks using a "cookbook" approach, which allows you to simply copy the procedure to get a satisfactory result or modify it according to your individual needs.

In addition to reading this book, keep an eye on its companion website, which features blog posts, links, videos, and discussions related to this book. Web links will be frequently updated there and new ones added as new software is released.

This book's companion sites can be found here:

www.sketchupfordesign.com—Author's companion site.

www.wiley.com/go/schreyer—Companion site for students and instructors.

3D for All

Because SketchUp is not domain-specific, it has found a following with professionals and enthusiasts from many disciplines. This is why you will find SketchUp mentioned in discussions not only by architects, landscape architects, urban planners, engineers, construction professionals, woodworkers, and timber framers but also by robot builders, artists, sculptors, model-plane builders, paper-plane builders, mapmakers, historians, 3D game developers, and movie set designers (just to mention a few).

The techniques in this book are applicable to a variety of disciplines. Although many examples come from architecture or construction, some are from other disciplines (e.g., landscape design and interior design). Whatever your background is, feel free to take the examples that are presented here and adapt them to your discipline. (See **Figure 1.5** for a non-traditional use of SketchUp). The techniques you learn will be equally useful.

Taking this one step further, I can even say that I can't think of anyone who should *not* be using SketchUp. Living in a three-dimensional world and dealing with three-dimensional objects, everyone has the need at some point to model and visualize in three-dimensional space.

Consider this hypothetical situation: You want to build a deck in your backyard and need to explain to the builder how it should look. Another example is a physicist who needs to explain a lab setup in a presentation. Far too often we resort to 2D representations of our thoughts (the classic floor plan, for example), which leaves too much unknown and unexplorable.

Admittedly, many people are not trained in hand-sketching in 3D, which often leads to poor visualizations of things that can actually be quite interesting when presented right. *That is where SketchUp shines.* Its 3D modeling capabilities and its ease of use make it a simple yet very powerful tool for anyone to give shape to their thoughts.

Figure 1.5: Rendered 3D QR-code model

How Does SketchUp Fit into the Designer's Toolbox?

As a student or professional, you likely already have assembled a sizable software "tool chest" by now. Depending on your discipline, this might include office software, CAD software, image-editing software, print layout software, analysis software (for energy or structural analysis, for example), coordination tools, and many others.

The free version of SketchUp fits into this collection very well. Depending on your needs and knowledge of the software, you can use it as an early design tool—after all, as its name indicates, it was developed for 3D "sketching." You can also use it for the complete design process from initial stages to a finished product (whatever that may be). SketchUp Pro broadens this even further by providing layout and documentation abilities and other professional-oriented tools.

SketchUp works well with other software. 3D models from SketchUp can often be directly opened in other software, making data exchange easy. Even if that isn't available, SketchUp's built-in file exchange options allow you to export a 3D model in a variety of formats.

If SketchUp is already part of your tool set, then it is the best use of your time to expand on the skill set that you have developed and deepen your knowledge of this software. This book provides you many avenues to do so.

Windows or Mac, Free or Pro?

SketchUp comes in two flavors: free and Pro. It is also multiplatform software, which means it is available for both Windows and Mac computers.

In the free version, a user can do almost everything that is available in the Pro version. The main differences are that the free version does not include the more professionally oriented DWG/DXF file exchange options (plus some others), it also does not include the ability to create Dynamic Components (and report them), and it does not include Pro's excellent Solid Tools.

Looking at the Pro version, you will find that it comes with two additional pieces of software: LayOut, which is a tool for drawing preparation and presentations based on SketchUp models, and Style Builder, a program that lets you make your own hand-drawn styles based on pencil-sketched lines.

While the free version can be downloaded by anyone from SketchUp's website, the Pro version is moderately priced (under $500 in the U.S.) and can be purchased from the website as well. Both versions are currently available in twelve languages (Traditional Chinese, Simplified Chinese, Dutch, English, French, German, Italian, Japanese, Korean, Brazilian Portuguese, Spanish, and Russian), which gives this software global reach. At this point, qualified students (in the U.S.) can get a time-limited license to use SketchUp Pro for $45.

Depending on your needs, you have to decide which version is right for you. For almost all of this book's content, it is *not* necessary to have the Pro version; photorealistic rendering, many plugins, and scripting work perfectly well in the free version. This book, therefore, offers a cost-efficient entry into relevant and current topics (such as 3D modeling, rendering, and computational geometry). Because SketchUp comes in a free version, it provides an opportunity to use advanced software approaches without having to resort to costly software.

Nevertheless, some Pro tools are covered in this book (e.g., creating a Dynamic Component and using Solid Tools). Whenever a chapter in this book mentions a Pro tool, it is visually presented as a "Pro Only" section.

PRO ONLY

Pro Only sections look like this.

While all of this book's illustrations have been created using the Windows version of SketchUp, the tasks and tutorials are similarly usable with the Mac version. Menus and dialogs generally look the same and are in the same location on both platforms. There are minor user-interface differences, but those are easy to figure out. Consult SketchUp's help system if you run into trouble.

About SketchUp's Transition from Google to Trimble

On April 26th, 2012, Google (who had bought the SketchUp software in 2006 from @Last, its original makers) announced that they sold SketchUp to Trimble, a company known for AEC (architecture, engineering, and construction) technology and software. The announcement came with assurances from both companies that SketchUp will remain available in a free version as well as a Pro version and that software development will increase in the future. Given the transition to Trimble, some changes have to be expected, but for a foreseeable time, the largest extent of those changes will be differing URLs and documentation changes.

Fortunate for the user community, this transition infuses energy into SketchUp development and it is very likely that new features will be created soon while its core functionality (as described in this book) will be retained.

Keep an eye on the companion website (**www.sketchupfordesign.com**) during this period of transition. I will post updated URLs and announcements of new features as they become public. In the meantime, this book will use currently known URLs and the "Google" name where relevant to current installations.

It is a good idea to stay up to date with SketchUp. In addition to this book's companion website, bookmark the following sites to help you with this:

www.sketchup.com—The official home of SketchUp. You can download the latest free version or buy the Pro version here.

http://sketchupdate.blogspot.com—The official SketchUp blog—a great source for updates, tutorials, and tips.

http://support.google.com/sketchup—SketchUp's help system. If you get stuck, go here first!

www.alexschreyer.net—My personal blog where I frequently post about SketchUp and other AEC software.

@alexschreyer and **@sketchupplugins**—My Twitter handles under which I post news and links about SketchUp and AEC software.

How This Book Works

One way to use this book is linearly as a learning tool by moving from chapter to chapter. This method builds your skill set gradually and allows you to logically approach each subject.

You may also want to use it as a desk reference, or you might be interested just in individual chapters. In these cases, make use of the index and the appendices.

Some conventions in this book:

■ Whenever I mention a "window" (e.g., the Materials window), this means the dialog window that can be accessed from SketchUp's Window menu.

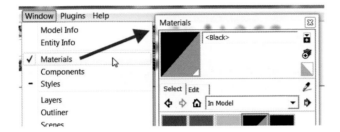

■ Other dialogs that open when the user clicks on something are commonly called "dialog" in the text.

■ Menu locations are typically presented in this format: **File → Open . . .**

■ Any toolbars mentioned in the text can be opened from the **View → Toolbars** menu in SketchUp. Plugins often install their own toolbars. Those will, of course, not be available until a plugin has been installed.

■ Following are some Mac-specific differences:

  ■ SketchUp's preferences cannot be found under the Window menu item, but instead are under the SketchUp menu.

  ■ Toolbars are called "Tool Palettes."

  ■ Instead of right-clicking to bring up the context-menu, you can left-click the mouse while holding the Control key.

Let's Go!

It's time to explore the world in the third dimension. Enjoy your modeling endeavors!

Chapter 2
A SketchUp Refresher

This chapter reviews some of SketchUp's basic techniques. You'll also learn about customizing the software environment and adjusting settings to help you with your daily tasks.

Key Topics:

- Where to get help with SketchUp
- Program interface and workspace customization
- Program and model preferences
- Working with templates
- Navigating SketchUp's 3D workspace
- Aids for accurate modeling
- Groups and components
- Applying materials and using other common tools
- Best practices for working with SketchUp

Let's Get Started!

Before we look at any of the more advanced SketchUp techniques such as plugins, rendering, or even scripting, it should be helpful for any user to review in a short chapter some of the basic modeling techniques. This chapter therefore presents in a condensed fashion an overview of the interface and the modeling and editing tools, as well as best practices and usability hints.

If you have no prior experience with SketchUp or would like to learn more about any of the basic topics, then my best recommendation is to get a copy of Aidan Chopra's excellent SketchUp primer *Google SketchUp 8 for Dummies* (Wiley Publishing). It is a thorough reference not only for modeling with SketchUp but also for creating building models for the free 3D world viewer Google Earth.

Two further sources for help are built right into SketchUp. One is the official help documentation. You can access it through the Help menu item, which simply opens a browser window at the following URL (if you end up using it frequently, it might be a good idea to bookmark it in your browser):

> **http://sketchup.google.com/support**—SketchUp's online help system.

Alternatively, you can turn on the Instructor feature by selecting Instructor from the Window menu. This opens (and keeps open) a window that displays a help page for the tool that is currently in use. For example, if you start using the Circle tool (by clicking on the Circle tool-bar button, selecting Circle on the Draw menu, or simply hitting the C key on the keyboard), the Instructor shows the appropriate help, complete with a small animation. (See **Figure 2.1**)

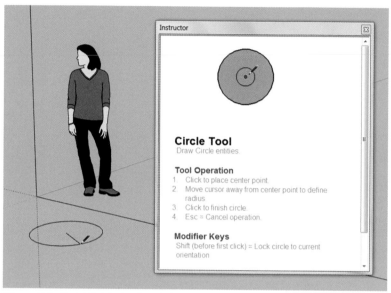

Figure 2.1: SketchUp's Instructor window

I also encourage you to visit this book's companion website (www.sketchupfordesign.com) for an updated list of links to online resources. In addition, there are currently many discussion forums, wikis, and online video sites available that offer help for both beginners and advanced users. The most popular ones are:

> **http://productforums.google.com/forum/#!forum/sketchup**—Sketchup's official help forum.
>
> **http://forums.sketchucation.com**—The SketchUcation forums, a user-based world-wide community.
>
> **www.aidanchopra.com**—Aidan's companion site to his book, which contains many YouTube instructional videos.

Interface and Program Setup

Once installed, SketchUp has a clean and rather empty appearance. The main portion of the screen is taken up by a large 3D-space work area in which only the ground plane (shown in solid green), sky (blue gradient), the three main axes, and a modeled person (included as a size reference) are visible.

The person in SketchUp's default template has historically been an employee from the SketchUp "crew." For version 8, this is "Susan." Earlier versions featured "Sang" and "Brad."

Above the work area, the main toolbar buttons are nicely arranged on a toolbar called Getting Started. **Figure 2.2** shows the workspace on Windows; **Figure 2.3** shows the Mac equivalent.

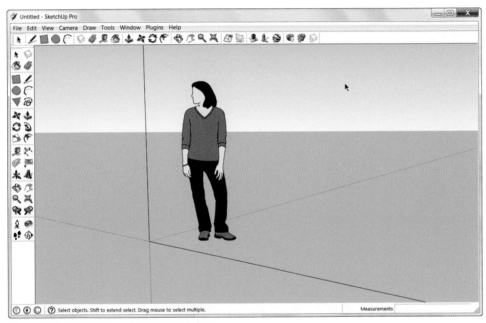

Figure 2.2: SketchUp's workspace on Windows

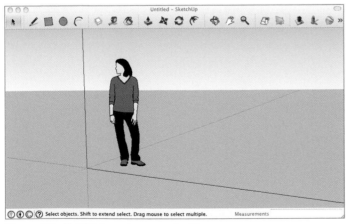

Figure 2.3: SketchUp's workspace on the Mac

SketchUp's numerous features can be accessed either through the screen menu, by clicking a button on the toolbars, or with keyboard shortcuts. As with many other programs, any toolbars that are not currently activated and visible can be displayed easily by selecting them from the **View** → **Toolbars** menu. You can see, for example, that an additional toolbar has already been activated in **Figure 2.4**. The toolbar called Large Tool Set has been activated and placed on the left side of the screen. It is always a good idea to have at least this one opened—even if no others are needed. In addition to the features in the standard Getting Started toolbar, it gives direct access to more navigation tools. However, it does not contain any of the 3D Warehouse or Google Earth links. If you need those, make sure you activate the Google toolbar.

It is a good idea to develop a well-organized workspace in any software. In SketchUp, this means that if you find yourself using the same tools over and over (and you can't access them using keyboard shortcuts, or you prefer toolbars), then you might want to display the appropriate toolbar and dock it on the screen in a convenient position. When you have arranged your workspace to your liking, don't forget to click on **Save Toolbar Positions** in the **View** → **Toolbars** menu so that you can restore them if a problem occurs. (SketchUp has been known to occasionally—and out of the blue—rearrange your toolbars.)

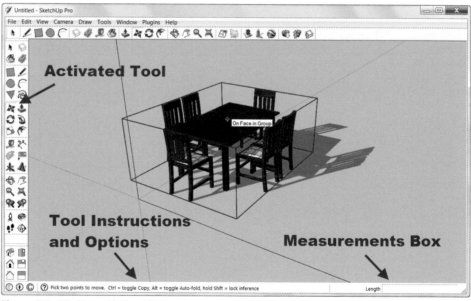

Figure 2.4: Working with SketchUp's tools

A very useful feature of SketchUp's user interface is keyboard shortcuts. Even with the default settings, many tools can be accessed by pressing single keys. For example, the "L" key brings up the Line tool, "R" the Rectangle tool, and the spacebar the Select tool (and cancels any other currently active tool). An overview of the default keyboard shortcuts (and most tools) is given in Appendix A. As you can see in **Figure 2.5**, you can, in addition, assign keyboard shortcuts to any tool that is available in the menu structure. Just remember to export your keyboard settings (from within the Preferences dialog) so that you have a backup available in case you need to reinstall SketchUp.

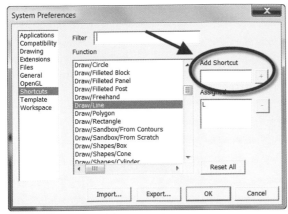

Figure 2.5: Keyboard shortcut preferences

You can even assign a keyboard shortcut to plugin-supplied tools (as long as those tools have created their own menu items)!

Adjusting Preferences

It is a good idea to spend some time tweaking all the options in SketchUp's Preferences to your liking, especially if you are a frequent user. The following is a list of the main options (access the dialog under **Window** → **Preferences**), along with some tips:

- **Applications**—Select the path to your default image editor here (e.g., Adobe Photoshop). This makes it possible, for example, to tweak textures using that editor by opening images in the external editor directly from within SketchUp.

- **Compatibility**—You can change your highlighting and mouse wheel preferences here.

- **Drawing**—You can select your click preferences here. For example, you have the choice of whether you want to draw a line by clicking and dragging or by clicking two points. Another useful feature is "Display crosshairs," which always gives you an axis reference when you have a tool activated.

- **Extensions**—This shows a list of installed extensions (also called "plugins"). The default installation should display the Sandbox Tools together with a handful of others. You can install plugins here using the "Install Extension..." button. It is also possible to disable and enable installed plugins here by simply unchecking or checking the box next to the plugin's name (a restart is often required). This sometimes becomes necessary when a plugin leads to crashes but you don't want to uninstall it by removing its files. You will learn more about installing, using, and uninstalling plugins in Chapter 4.

- **Files**—You can select locations on your hard disk for models, components, materials, and so on. These are set up by default in the main SketchUp installation directory, but sometimes it is quite convenient to also have a separate location where custom components are stored. A use case for this is a company in which several SketchUp installations need to access models located in central network storage (for details, see the note that follows this list).

- **General**—These are some miscellaneous settings of which the most important ones are "Create backup" (make sure this is checked in order to have a backup in case of a crash), "Auto-save" (set this to a short interval, such as every ten minutes), and "Automatically check for updates" (to be reminded when an update for SketchUp is available).

TIP

If you set up SketchUp to save backups, then there will be an SK**B** backup file saved in the same location where your SK**P** SketchUp file is saved. In case your main file becomes unusable, you can open the backup file directly with SketchUp. Just make sure you can see all files in the Open dialog by selecting "All files (*.*)" at the bottom of the dialog.

- **OpenGL**—These settings determine how the application graphics are being processed, which in turn determines how responsive SketchUp will be when you work with it. Unless you have a very old graphics card, make sure "Use hardware acceleration" is checked. This lets the graphics card do most of the display rendering and results in a responsive 3D work environment. Also check the "Capabilities" table below these settings. It shows which capabilities of your graphics card have been found by SketchUp. It is always a good idea to select one of the rows where the anti-alias number is larger than zero so that lines on-screen don't appear jagged. Experiment with these settings until you find the combination that works best for you.

- **Shortcuts**—As mentioned earlier, this allows you to add your own keyboard shortcuts.

- **Template**—You can select your default template here. To efficiently use a template in SketchUp, modify an empty file to your liking (including all settings in **Window → Model Info**, since they will be stored in the SketchUp file). You can even preload components and materials. Then save the file as a template (choose **Save As Template. . .** under the **File** menu) and preselect the file in this preference tab.

- **Workspace**—This allows you to change the size of the toolbar buttons and to reset the workspace.

IN-DEPTH

Using SketchUp on Multiple Computers

If you use SketchUp on more than one computer, it might make sense to work from the same set of components, materials, textures, and so on. You can then, for example, create a collection of your own (or your company's) custom components and materials that is synchronized across all the computers on which SketchUp is installed.

To accomplish this, there are two options:

- **Work with a central repository.** This option requires that you set up central data storage (e.g., network-attached storage or a server), where all files can be stored. Then all you need to do is go into SketchUp's Preferences dialog and set all file locations to the appropriate folders on the network location. A drawback to this approach is that in order to access the stored data you need network access. Depending on your network connection and the size of your models, loading and saving might be slow.

- **Work with a synchronized repository.** This option is faster and safer because the repository is synchronized across all computers. It also ensures that the repository is available even when there is no network connection—for example, when you take your laptop on the road. The easiest way to make this option work is to sign up with Dropbox (**www.dropbox.com**) or a similar service. Then download the client application for your system from Dropbox and set up a dedicated folder on your computer to hold all synchronized data (you may use it for more than just CAD files, of course). After you do this on all your computers that run SketchUp, go to SketchUp's preferences and add the local (on your computer!) synchronized folder to the Files options. Thereafter, whenever you save, for example, a new component, Dropbox synchronizes it across all linked computers and, thus, it will be available wherever you work with SketchUp.

With any of these methods, make sure you set up a good check-in/checkout system or use your network's document storage provider's history features so that a user can't inadvertently overwrite somebody else's files.

Working with Templates

Beyond the program's settings, there are quite a few settings, data, and other items that are stored within a SketchUp file (with an SKP extension) and therefore can be added to a template. This file is opened as the default empty file each time SketchUp starts and includes the following:

- All settings in the Model Info dialog and therefore (among others) dimension styles, text styles, and units
- Any entity placed in the 3D workspace (e.g., the person reference in the default template)
- The model's geo-location
- Model credits
- Animation tabs and their views with all settings, such as View styles
- In-model components, materials, styles, layers, and so on
- Component attributes
- Plugin data if a plugin stores data within the file

As you can see from this list, many options might warrant tweaking and saving in a template. Also, because the template settings can't be applied retroactively to an already existing file, it makes sense to preset template options before starting work in a file. Let's look at setting some of the most useful options.

Setting Units and Fonts

Figure 2.6 shows the Units settings. Depending on your profession or where you are in the world, you may want to select architectural units (e.g., **6′ 5 ½″**), decimal units (inches or feet or a metric unit—e.g., **2.5′** or **3.2 m**), engineering units (e.g., **100.623′**), or, simply, fractional inches (e.g., **250 ⅝″**).

Depending on the resolution of your work (are you using SketchUp for woodworking or to design houses?), make sure you preset both the length snapping and the angle snapping to values that make sense for you. Finally, set the precision to display a number of digits or a fraction that is appropriate for your work. Whatever you set here, rest assured that the software internally stores all numbers at a much higher degree of precision.

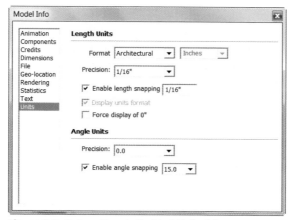

Figure 2.6: Units settings in the Model Info dialog

A great way to personalize your SketchUp output is by using a custom font for in-model text as well as leaders and dimensions. You can do that in the respective settings (see **Figure 2.7** for text setting options and **Figure 2.8** for dimension settings).

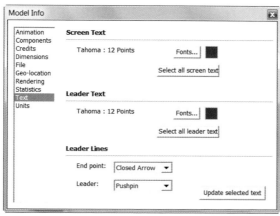

Figure 2.7: Text settings in the Model Info dialog

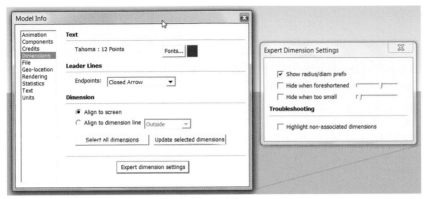

Figure 2.8: Dimension settings in the Model Info dialog

Figure 2.9 shows a sample of various fonts: Windows' standard Tahoma font as well as some custom technical fonts and a handwritten font. You can even create your own font using your own handwriting. Some Web resources are mentioned as follows.

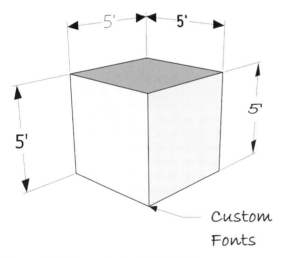

Figure 2.9: Using custom fonts in SketchUp

www.yourfonts.com—You can make your own font using this Web service. Alternatively, if you use a tablet device, look for software that allows you to do this by sketching the letters (Windows Tablet PCs come with such software preinstalled).

www.dafont.com—This is only one of many websites that offer free and for-sale fonts. To install fonts, first download the desired font file (which can have any of these formats: TTF, OTF, or FON). Then, depending on your computer setup, either double-click the file and select Install or drop the file into the operating system's Font folder (usually under Preferences or Control Panel).

Adding Components

Think carefully which components you want to add to your SketchUp model because each one will increase the SketchUp file size. Having said that, it is always possible to purge unused file-space-heavy components and materials by using the options in the respective windows (see **Figure 2.10** for the Components window);therefore, it might be a good idea to start with a larger file and then simply get rid of unused components later.

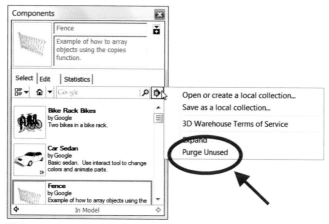

Figure 2.10: Purging unused components

Samples for preloaded components in architecture are people cutouts, a minimal set of trees and shrubbery, as well as any annotation elements that you might want to add to your model (e.g., a "north" arrow). While SketchUp's default components are easily accessible through the component windows and therefore should not need to be loaded into a template, it's a good idea to do this if you have created your own set of these components to give your models a more personal touch.

One good example for this is your own person cutout (akin to the "Susan" figure that shipped with version 8 of SketchUp). One of SketchUp's common criticisms is that its output can easily be identified as coming from SketchUp because this person is very often left in the view.

To make your own person cutout, load a photo that you like into the view, scale it to the correct size, and then trace the shape of the person (and delete the outside faces). Then just apply color to all faces and make it a component. Make sure you select "Always face camera" in the Create Component dialog. Instructions for creating components appear later in this chapter and a similar example (creating a cutout tree) is included in Chapter 5.

Views

If you often switch between working in perspective mode and in parallel projection views (top view, front view, etc.), then it might make sense for you to preset the most important views as animation tabs. To do this, create new tabs through the **View** menu by clicking on **Animation → Add Scene** once you change all view settings to your liking (you can also update a tab's settings by right-clicking on it and selecting Update). You can switch between parallel and perspective views and select standard views from the Camera menu.

In **Figure 2.11**, the tab names have been changed in the Scenes window and "Include in animation" has been unchecked so that these views will be excluded from any animations that you might create using the tabbed animation feature.

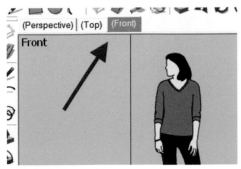

Figure 2.11: Several preset view tabs

Completed Template

Figure 2.12 shows a cleaner version of SketchUp's default template with several customizations. Feel free to create either a subtle or a bold template for your own work. Either way, visual consistency goes a long way when you use SketchUp professionally.

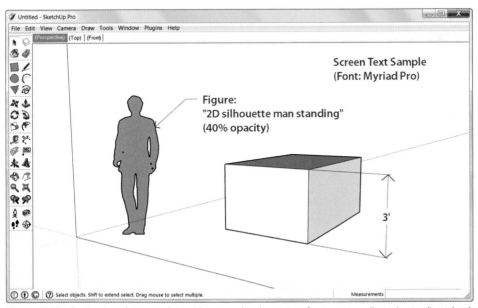

Figure 2.12: Custom template complete with pre-made tabs, custom fonts, custom dimensions, adjusted styles, and architectural units

SketchUp's Tool Set

Navigating the 3D Model

SketchUp provides various means to move around 3D space while you are working on your model. Analogous to many other CAD software programs, there are tools for zoom, pan, and orbit. You can access these using the buttons on the main toolbar. If you are using a wheel mouse, however, you can do this much faster by rotating the wheel (to zoom in or out), pushing the wheel and dragging the mouse (to orbit), or holding the Shift key while you push the wheel (to pan sideways).

Two further tools of interest in this context are the Walk and Look Around tools:

- **Walk**—This tool allows you to explore the 3D space interactively by walking back and forth through space as if you were walking through a building, for example. To use the tool, click on the icon that shows the two footsteps, which is found on the Large Tool Set toolbar. Then use your mouse (click and drag) to navigate.

- **Look Around**—As the name implies, the way this tool works is similar to the way you move your head around to explore a space.

IN-DEPTH

Alternatives for Controlling SketchUp's View

The aforementioned built-in methods are not the only tools that you can use to orbit, move, or walk around. Here are some suggestions for alternatives.

One method is a 3D mouse (e.g., the SpaceNavigator made by 3D Connexion), which behaves more like a joystick and allows you to move more intuitively through 3D space because it combines the zoom, pan, and orbit motions into one tool. A common use case for this tool is to position it at the left hand (if you are right-handed) and use it in combination with the regular mouse at the right hand. (See **Figure 2.13**.)

Another option is to use a device that can be operated in 3D space similar to the actual motion in SketchUp. This offers the possibility of experiencing SketchUp's 3D space in a more immersive fashion. An example is a Wiimote controller (taken from a Wii video game console). This tool is very inexpensive ($40 is a current retail price), and its sensors (an accelerometer and others) can be set up to control any software using the freely available

GlovePIE software. (See **Figure 2.14.**) If you want to try this out for yourself, follow the instructions on my website: **www.alexschreyer.net/cad/when-wii-met-sketchup**.

Figure 2.13: Workspace with 3D mouse

Figure 2.14: Using a Wii remote to move around in SketchUp

Both of these tools might be better suited to exploring 3D space and models rather than being part of actual 3D modeling. Thus, they are very useful for presentations, client walk-throughs, and the like. Feel free to experiment with them (and others) and see where they might fit into your workflow.

Accurate Modeling

The often-voiced notion that SketchUp is capable of merely "rough" and sketchy work is quickly dispelled when we look at how the program stores numbers internally. For example, querying the value of the mathematical constant pi using the internal scripting tools (covered in detail in Chapter 6) gives us the number 3.14159265358979. As you can see, this number is stored internally with precision to 15 digits. Any length or position value in SketchUp's 3D space is therefore internally stored with the same degree of precision.

TIP

Independent of where you use SketchUp in the world and what your unit settings are, length and coordinates are always stored internally as inches. These values are then conveniently converted into your local unit format before they are displayed in the Measurements Box (the box that is displayed at the bottom of your screen by default in Windows) or anywhere else in the program.

To take advantage of this level of precision, use any of these tools when you are modeling:

- **Object snapping**—Like other CAD software, SketchUp offers a complete set of object snaps. This allows you to snap to endpoints, midpoints, edges, circle centers, and other locations. (See **Figure 2.15**.)

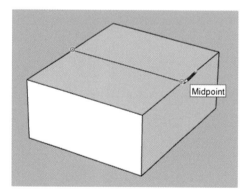

Figure 2.15: Object snapping while drawing a line on a box

- **Length snapping**—When modeling a line, for example, the mouse cursor snaps in increments of length that you have preset in the Preferences dialog (e.g., at every 1/8"). You can then simply click once the correct value has been reached, and the line will have the exact length. (See **Figure 2.16**.)
- **Inferences**—One of SketchUp's main strengths is a very intuitive inferencing system. This means that you can reference a point in 3D and model relative to it. To use this feature, start a tool (e.g., a line) and then move around 3D space. Especially if you move parallel to one of the axes (your "rubber band" line will change to the color of the axis), you will occasionally see a dashed line and a temporary reference point shown as a dot to

indicate that you have acquired a reference point. The temporary tooltip will also display "From Point" to indicate that this has happened. You can then simply click, and the click position will have been determined by the referenced point's position.

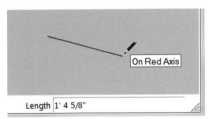

Figure 2.16: Line length snaps with preset 1/8″ increments

It sometimes helps to constrain the direction of the temporary line. This is done by moving along the intended direction (e.g., parallel to an axis) and then holding down the Shift key. The rubber band line will increase in width to indicate that a temporary constraint has been activated. (See **Figure 2.17**.)

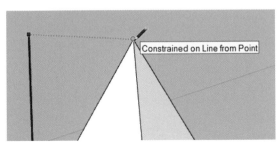

Figure 2.17: Using a constraint on the blue axis together with inferencing to acquire the pyramid's height as the top point for a vertical line

- **Direct value entry**—Many tools provide the option to enter values directly into the Measurements Box—even without the need to activate it by clicking. For example, you can start creating a line with a click, then reference one of the axes and just type the intended length for the line (e.g., 10′).

SketchUp makes it easy to work with multiple unit systems. You can enter any unit into the Measurements Box (e.g., *30cm*), and SketchUp internally converts it into your preset unit system and uses the accurate values. You can even enter mixed units (e.g., *30cm,3′* for the dimensions of a rectangle; see **Figure 2.18**).

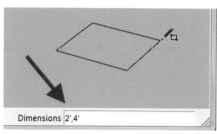

Figure 2.18: Direct value entry for a rectangle

Example 2.1: Starting a Trellis

Let's review and practice these techniques with a simple exercise: Let's build a garden trellis. As a first step in our trellis-building exercise, we need to create two vertical 8'-tall 6 × 6 posts.

1. Create a 5.5" × 5.5" square (using the Rectangle tool) on the ground plane (orbit your view if necessary so that you are looking "down" onto the ground). Use direct value entry into the Measurements Box for this:

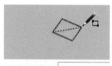

2. Then use the Push/Pull tool to pull this rectangle up to an 8' height:

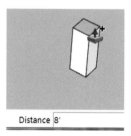

3. Now use point inference on the front corner of the newly created post to position the new rectangle a distance along the red axis relative to the first post. Create the base square as you did in step 1.

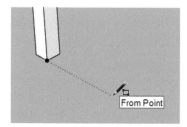

4. Use Push/Pull again, but this time don't enter a length value; rather, use inference with the top of the first post to make the second post exactly as tall as the first.

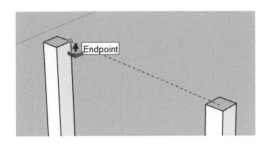

Temporary References

Beyond the aforementioned capabilities, SketchUp offers temporary lines and reference points for accurate modeling. These are created using the Measure and Protractor tools. Let's review their functions by continuing with our trellis model.

Example 2.2: Creating Beams for the Trellis

1. Start by aligning your view with a vertical plane similar to the image shown here. This allows you to create a vertical rectangle that you will take as the basis for a 2 × 6 beam. Then draw the rectangle using the accurate dimensions (1.5″ × 5.5″).

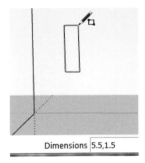

2. Now use the Push/Pull tool to extrude it to 8′.

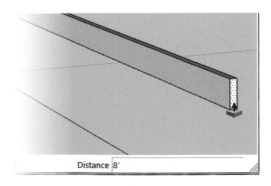

3. It is a good idea to align your view with the side of the beam—either orbit and zoom or right-click on the side of the beam and select **Align View** from the context menu.

4. Start sketching temporary lines on the side of the beam. Using the Tape Measure tool, click on the top edge (make sure it snaps to the edge and not a point like the endpoint), and move the mouse downward. You should now see a horizontal (dashed) guideline. Click when you reach 2″ (this is possible because you preset length snapping earlier), or enter *2* into the Measurements Box.

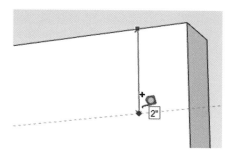

5. Now click at the point where the guideline intersects the right edge. Because you picked a point and not an edge as the start, you will not create a parallel guideline but rather a guide point. Move left along the guideline and click at 3″ to place the point.

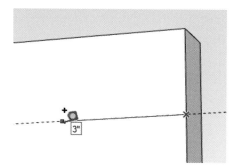

6. Use the guide point as starting point for an angled guideline, which can be created with the Protractor tool. After starting the Protractor tool, you first need to place the center of the tool gizmo (the round protractor icon). Click on the temporary guide point for this. Next, you select a zero reference (click on the horizontal guideline). Finally, move down to the left until the protractor snaps at 30 degrees. Then click to place the angled guideline.

TIP

You can constrain the protractor gizmo to a particular plane by hovering over a face that is parallel to that plane and then pressing the Shift key. As long as the Shift key is pressed, the gizmo will not automatically shift its orientation based on a face's orientation below the mouse cursor.

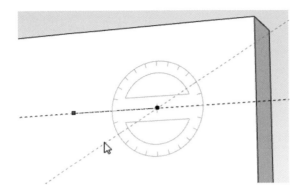

7. We can use the guidelines to "score" the front face by drawing lines onto them (make sure you snap to endpoints and intersections) and using the Push/Pull tool to remove the bottom corner of the beam (push all the way to the other side of the beam to completely remove the bottom).

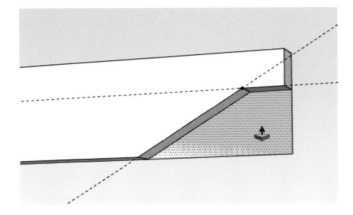

8. You can now erase the guidelines either in the same way as you would any other object using the Eraser tool, or by going to the **Edit** menu and selecting **Delete Guides** (this will erase all guidelines and points in the entire model, though).

Groups and Components

Groups and components are at the heart of modeling with SketchUp. Although both have one thing in common—they combine edges and faces into one object and thereby provide organization for the entire model—they work differently and are being used for different purposes. In essence, these are the differences:

- **Group**—When geometry becomes grouped, all selected entities (edges, faces, and other groups or components) are combined and are then selectable as one object. Any geometry that is drawn on top of a group does not "stick" to the group. You can also apply modifications such as scaling, moving, or copying to the entire group. It is important to note, though, that copying a group copies all of its contents and therefore doubles the storage space required in the SketchUp file.

■ **Component**—A component is also a collection of geometry and other groups/components. Unlike a group, a component that is placed in the model is always just an insertion (an *instance*) of a component definition. A good way to imagine this might be that the geometry that makes up a component gets stored in a hidden part of the SketchUp file. When a component is placed into the model, a representation of it is inserted at the point you click. Multiple insertions (or copies) of the same component need only one insertion point per copy. While any of these insertions can have a different location, rotation, or scale, the underlying geometry is safely stored in that hidden location in the file. As a result, the file space required for a component is simply a single copy of the geometry plus a single insertion point for each copy. This makes components very efficient in terms of storage. In addition to this, components have their own (local) coordinate system that is stored within the component. They can also be set up so that they always stick to the horizontal or vertical plane or any face. Furthermore, they can have a property whereby they will always face the viewer (this is the case with the 2D person cutout that comes with the default template). As in the case of a window component, they can also be set up to cut a hole in the underlying geometry.

Beyond these differences, components can be shared in the 3D Warehouse, SketchUp's online 3D model storage service. And as of version 7, components can also be set up to have interactive behaviors and may contain data when they are set up as Dynamic Components (we explore these in more detail in Chapter 3). This makes components very flexible and allows them to be used in a multitude of ways, for example, as parametric objects.

So how are groups or components created? After selecting the edges, faces, and groups/components that will make up a new group/component, one option is to right-click and select the appropriate item from the context menu (see **Figure 2.19**). Another is to use the menu items in the Edit menu. The fastest method may be a keyboard shortcut—typing "G" by default combines all selected objects into a component. See **Figures 2.20** and **2.21** for examples of groups and components, respectively.

TIP

Unfortunately, the Make Group function has no assigned default keyboard shortcut. However, this can easily be remedied in the Preferences window, where any convenient shortcut (e.g., Ctrl+G) can be assigned to the function.

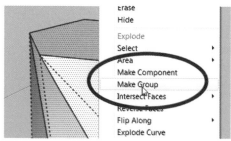

Figure 2.19: Creating a group or component using the context menu

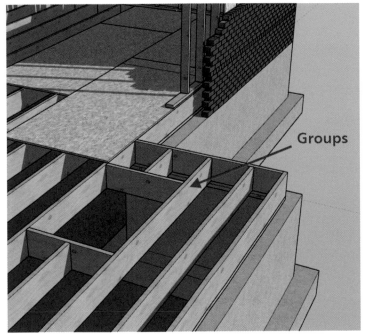

Figure 2.20: Groups are best used when there are objects of many different sizes and configurations (especially if their texture does not tile)

Figure 2.21: Components are best used when there are many identical objects

After inserting copies of components into your model, you can do this:

- **Edit the component definition.** (Double-click one insertion to get into edit mode.) All copies of the same component will automatically be modified when you edit something.

- **Make a component instance unique.** After inserting many copies of a component, all insertions ("instances") are representations of that component. If you need one of them to be different, then you can click on **Make Unique** in the context menu. This creates a copy of the component definition and replaces the selection with it. This component will now behave differently from the original definition.

- **Replace all copies of a component.** This is useful if you want to use a placeholder component in your model and replace it later or if clients change their minds and you need to replace one set of objects with another. This is accomplished by going into the Component window and displaying all of the components in the current model (click on the house symbol). You can then right-click on the to-be-replaced components and pick **Select Instances**. Then browse to the component with which you want to replace these, right-click on the component again, and select **Replace Selected**.

Example 2.3: Using Components for the Trellis

Let's practice working with components by continuing with our trellis exercise:

1. Because the trellis beams will have the same cuts applied to both ends, it is useful to create a nested component. To do this, let's first reduce the beam length from 8′ to 4′. Use the Push/Pull tool for this.

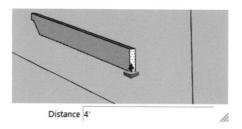

Distance 4′

2. Now highlight the beam by either triple-clicking one of its sides or selecting it with a window selection. Then press the "G" key on the keyboard. A dialog will come up that allows you to give the new component a name. Keep all other settings as the default.

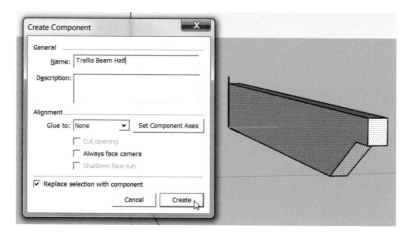

3. Use the Move tool to create a copy (hit the Control key after activating the Move tool to toggle copying). After creating the copy, hover over one of the rotation icons on the top of the beam to rotate the copy about the vertical axis.

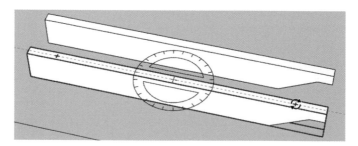

4. Move the beams back together, highlight both, and create a new component from this.

5. Because four identical posts will be holding up the beams, it is a good idea to go back to the posts you modeled in the earlier exercise, delete the second one, and turn one of them into a component that we will name "Post."

6. At this point, you can start moving the trellis beams into position. We need two support beams that will need to be attached 5½" from the top of the posts.

TIP

When you move the beams into position using the Move tool, use object snapping (mostly on endpoints) and direct distance entry to your advantage to place the beams exactly where they need to be. Pick up objects at a snap location and drop them at another. Never just "eyeball" pick points.

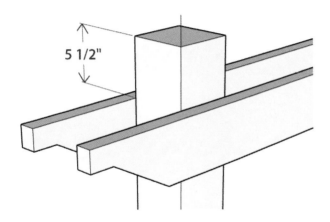

7. After the two beams have been placed, make a copy of the entire portal, which you will use later for the other side.

8. You can now take one of the beams and place a copy perpendicular to the portals.

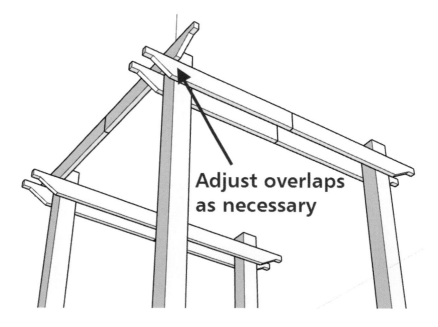

Adjust overlaps as necessary

9. Let's use the divide option of the copy function in this step to evenly distribute the cross-beams over the trellis top. To do this, select the last beam you placed, start the copy function (Move tool + Ctrl key toggle), and use inference to acquire the point in the next graphic as the base point for the copy. Then copy it over to the other post.

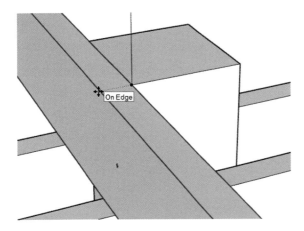

10. Without exiting the Move tool, type /5 on the keyboard (which will then appear in the Measurements Box). This evenly divides the space between the two copies into five segments and places copies of the beams at even distances. You can experiment with other spacings now until you exit the Move tool—at that point, the copy operation is final.

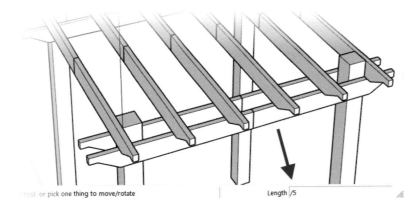

11. At this point, the trellis is complete. Let's use it, however, to see a benefit of having all beams be components, by applying an adjustment to the beams. To do this, double-click on one of the top beams—this gets you into component edit mode (the rest of the model will be grayed out).

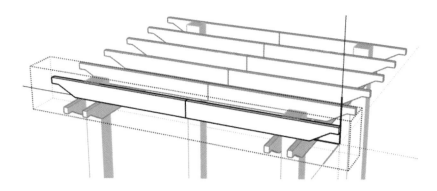

12. As you remember, this only got us into the beam component; we now have to double-click one of the half-beams to get into the underlying component.

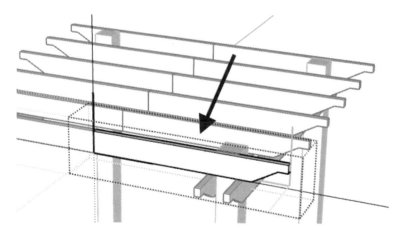

13. This is where we can make our modifications. Because all of the half-beams are instances of this component, all of them will be updated. To illustrate this, you will change the sloped edge into an arc edge. As before, use the Push/Pull tool to cut away the material at the bottom.

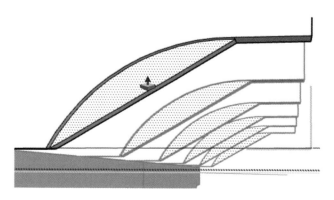

14. We can do further adjustments here—for example, we can stretch the beams to an overall length of 10′ (and thereby give them more overhang) by highlighting the entire end section and using the Move tool to move that geometry along the axis of the beam.

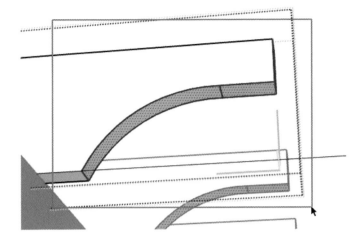

15. Finally, we don't want to see the edge at the center of the beams. Remember: We have this edge because we have two adjoin components at that point. Fortunately, SketchUp has an easy way to hide edges. Simply start the Eraser tool. If you hold the Shift key while you "erase" these edges, then you are hiding them instead.

TIP

Do not delete these edges. Whenever edges that bound a face get erased in SketchUp, all adjoining faces also get erased (because they need edges to exist).

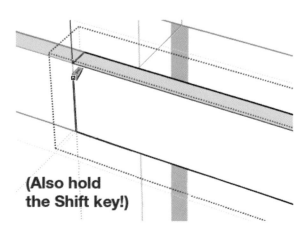

(Also hold the Shift key!)

16. Now click twice outside the actual model to get out of the component editing mode, and enjoy your work! This is what you should see:

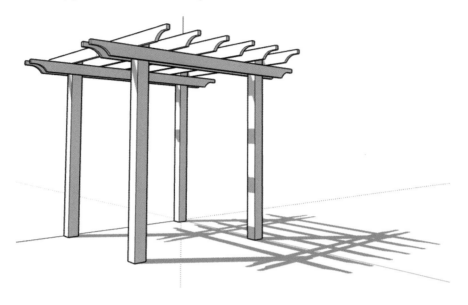

17. As a final step, take a look at the Components window. After opening it, click on the house symbol to see which components are loaded into your file. If you did everything right, you should see the three custom components in there: Post, Trellis Beam, and Trellis Beam Half (as well as the default Google person, of course). In this exercise, you practiced various techniques:

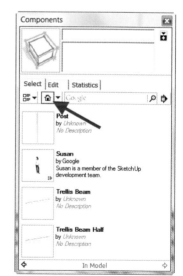

- Creating components

- Working with nested components

- Moving, copying, and rotating objects

- Accurate positioning using inferencing, object snap, and direct distance entry

- Divide-copy

- Editing components

I recommend you practice these techniques as much as possible. As you will see in the next chapters, a well-organized component-based model is essential to getting the most out of SketchUp.

Applying Materials

SketchUp has a very good and intuitive support for applying materials to faces. This is useful for a variety of activities, from a simple presentation to preparing a model for photorealistic rendering to modeling for Google Earth. Although SketchUp ships with a variety of good-looking materials (in the form of seamlessly repeating textures), it is important to note that you can always create a new material from your own textures. To illustrate this, let's apply a wood material to the trellis.

Example 2.4: Applying Materials to the Trellis

1. Open the Materials window and browse to the Wood materials. This gives you a choice of the materials that ship with SketchUp. Before we can apply a material, though, it is good practice to apply the material to the faces inside the component.

> **TIP**
>
> You can, alternatively, apply a material to an entire component simply by "painting" on it. However, if you do that, then you don't have as much control over positioning the texture as you have by applying it directly to faces.

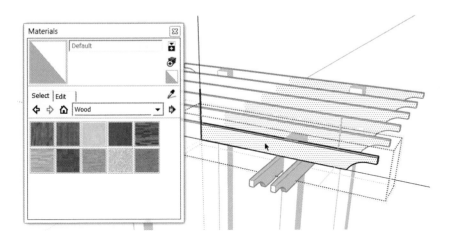

2. Although we have no plywood in our model, we'll use the Wood_Plywood_Knots material for our wood. Feel free to modify its appearance size on the Edit tab of the Materials window.

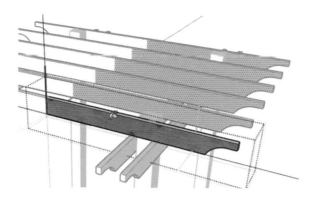

3. As when you previously edited the components, when you start applying materials in the component editor, all instances of that component will be updated. Therefore, after you finish applying the wood material to one half-beam, all of the beams are completed!

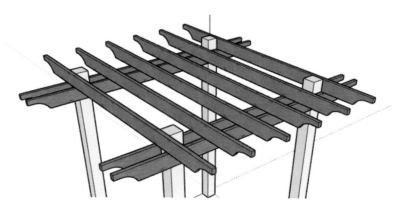

4. The next logical step is to go into editing mode for the posts and apply the same texture there. When you do that, though, you'll see that the material is oriented horizontally, which is not correct, as the wood grain runs along the posts and, therefore, vertically.

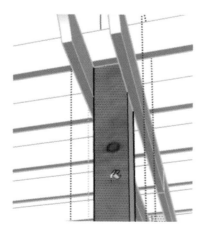

5. To correct this, we must rotate the texture. Fortunately, SketchUp has a very easy-to-use built-in system to modify a texture's orientation. To access it, right-click on the texture and select the context menu option **Texture → Position**. The screen will change to show the entire texture on a vertical plane along with four adjustment handles:

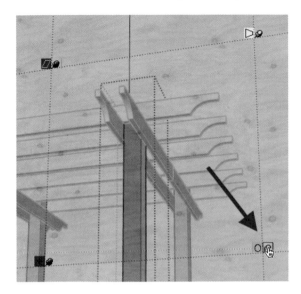

6. Now simply click and drag the scale/rotate handle (on the bottom right) and adjust the orientation of the texture. You can also experiment with the other handles to adjust position and skewness. Once you are done, right-click on the texture and click on **Done**.

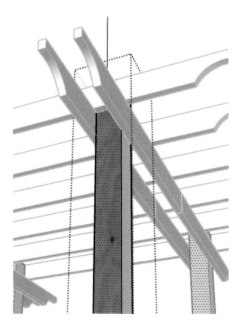

7. To apply this corrected texture to all the other sides of the post, use the eyedropper in the Materials window to pick up this texture from the front face of the post. Then simply apply it using the paint bucket on the other faces.

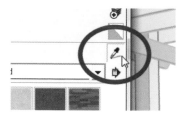

8. Now click outside of the editing area to get out of component editing mode. At this point, all faces of the trellis should have correctly oriented textures. Your trellis should now look very realistic:

In this exercise, we reviewed some basic concepts for placing materials on faces. This will again be useful in a later chapter when we look at photorealistic rendering of SketchUp models.

Other Tools

SketchUp has many more tools than could be covered in a one-chapter review. The following list gives an overview of the tools not reviewed in this chapter. Consult the help topics for further instructions on these.

- **More geometry tools: Circle, Arc, Polygon, Freehand**—These are best explored by experimentation. In each tool, keep an eye on the status bar and on the Measurements Box to see which options are available.

> Circles and arcs are always made up of line segments in SketchUp because SketchUp is a polygon-modeling software (it can work only with polygonal faces and edges). Therefore, remember to set the number of segments in the Measurements Box to an appropriate value—few segments make for "edgy" circles and many segments create rounder circles but increase file size.

- **The Modify tools: Rotate, Scale, Offset**—In many ways, the **Rotate** tool works similar to the Move tool. It allows you to rotate objects but also to make rotational copies (a good example is positioning chairs around a circular table). **Scale** allows scaling of any object either graphically or by entering a scale factor. **Offset** creates offset copies of edges on a face.

- **The very useful Follow-Me tool**—This tool is basically a sweep tool with which a face can be "extruded" along an edge (a set of lines and arcs). This tool can be used to make building foundations, roofs, and road curbs but also rotational objects like balls and cones. Experiment with it to understand its functionality.

- **Axes**—Sometimes it is convenient to reposition the coordinate system reference. This is usually the case when a model (e.g., a building) is rotated and not perfectly oriented in the north-south direction. This tool allows you to move the axis reference to any convenient orientation.

- **The Sandbox tools**—This set of tools resides on the Sandbox toolbar and is not activated by default (activate via the **View → Toolbars** menu). These tools allow you to work with terrain objects (i.e., triangulated surfaces). You can use these tools to create terrain, modify it, and add building footprints and roads.

- **Intersect and Solid Tools**—The Intersect tool is useful for easily finding intersections (intersecting faces and edges) between two objects in 3D. While it is not a true Boolean tool (due to the nature of SketchUp in not using actual solids), it permits you to find intersecting volumes and perform manual cleanup of geometry.

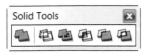

SketchUp's Solid Tools

The Solid Tools that were introduced with SketchUp 8 Pro (Intersect, Union, Subtract, Trim, Split, Outer Shell) are a more convenient way to use Boolean operations for modeling than the Intersect tool. They operate on grouped geometry that cleanly encloses a volume (often called *manifold* or *watertight* geometry).

- **Shadows**—SketchUp has an accurate shadow generator that allows shadows to be displayed in a model at any time and date and at any given geographic location. This is useful for solar shading studies and for adding realism to model views.

- **Photo match**—As SketchUp is used to model existing buildings for Google Earth and other software, several methods exist to create an accurate representation of a 3D object. One is the "ground up" method of modeling a building and then texturing it (using the methods shown in Example 2.4). Another one is the Photo match tool that uses a photo together with an adjustable axis to model a building by sketching on one or more photos.

- **Geo-location**—To be able to correctly place an object on the earth's surface (and even below the ocean!), SketchUp has a tool that allows you to pick a map-based location and insert a rough terrain model into a SketchUp model.

We will use some of these tools in later chapters. For now, if you are interested in more introductory and in-depth tutorials, review SketchUp's help files or refer to any of these excellent reference books:

- Brixius, Laurent, ed. *Google SketchUp Workshop.* Burlington, MA: Focal Press/Elsevier, 2010.
- *Catchup.* An online SketchUp magazine from SketchUcation. Available at: http://sketchu cation.com/catchup.
- Chopra, Aidan. *Google SketchUp 8 for Dummies.* Hoboken, NJ: Wiley Publishing, 2011.
- Chopra, Aidan, and Laura Town. *Introduction to Google SketchUp.* Hoboken, NJ: John Wiley & Sons, 2008.
- Fukai, Dennis. *3D Construction Modeling.* Gainesville, FL: Insitebuilders, 2004.
- Holzner, Steven. *Sams Teach Yourself Google SketchUp 8 in 10 Minutes.* Upper Saddle River, NJ: Pearson Education, 2011.
- Killen, Tim. *SketchUp Guide for Woodworkers* (eBook). Newtown, CT: Taunton Press, 2010.
- Roskes, Bonnie. *Google SketchUp Cookbook: Practical Recipes and Essential Techniques.* Sebastopol, CA: O'Reilly Media, 2009.
- *SketchUp-Ur-Space.* An online SketchUp magazine. Available at: www.sketchup-ur-space.com.
- Tal, Daniel. *Google SketchUp for Site Design.* Hoboken, NJ: John Wiley & Sons, 2009.

SketchUp Best Practices

The following is a list of best practices that you should always keep in mind when you are working with SketchUp:

- **Spend some time to make your own template and adjust the program's settings.** This was mentioned earlier in this chapter, and I can only reiterate how adjusting your workspace to your liking and comfort will help you when working with SketchUp (or any program, for that matter).
- **Save early and save often.** It is generally best to hit the Save button as soon as you start a new file and even before you start working on it. I have seen too many people lose hours of their work simply because they didn't do this. Once a file has been saved, SketchUp's autosave function keeps saving the file and making backups.
- **For computer-lab-based readers: Write your name on your USB stick.** It always amazes me how many USB memory sticks get left behind in computer labs or lost each semester. Therefore, if you work in a computer lab and you save your work to a USB stick, write your name and e-mail address on it; you might even put a file on it that lists your full contact data.
- **Group faces (or turn them into a component) as soon as they make up a logical object.** In SketchUp, all faces that touch each other become "sticky." This means that if you add a face on top of another face, you won't be able to remove the second face by simply moving it. A good practical example is that as soon as you have extruded a rectangle to form a 2×4 wood stud, turn it into a group or a component. That way, when

you make a copy and place two of them side by side, they will behave like two physically separate objects.

■ **Use components for repeating objects.** As soon as you have an object that will be copied multiple times in its current shape (e.g., floor joists or streetlights) or simply scaled, use components instead of groups. Components save file space and keep SketchUp working smoothly.

■ **Model as precisely as possible.** Use SketchUp's precision modeling tools to create a model that is accurate and doesn't contain "crooked geometry" or gaps. A common problem occurs when one corner of an otherwise planar face moves out of plane. At that point, SketchUp needs to triangulate the face, which in turn breaks it up and makes texturing and other modeling tasks harder (e.g., using the Push/Pull tool or moving edges). Try to avoid this by modeling orthogonally where appropriate and using dimension entry and object snaps wherever possible.

■ **Keep the polygon count low.** Because SketchUp is a polygon-based modeler, any curved shape must be approximated by a certain number of polygons (typically, triangles, rectangles, or polygons with more than four edges). The more polygons a model has, the more the graphics card will have to work to display the model and any shadows on it. This is true especially if you turn on edge display or even custom styles. A technique to reduce the polygon count is modeling circles or arcs with only the minimum number of edges needed. It is also beneficial to use components as much as possible, especially for curved objects that repeat.

■ **Use proxy components if needed.** If you need to work with high-polygon objects—for example, detailed trees for rendering—then insert *proxy components* (which can be as simple as a box with the rough outline shape of a tree) into your model while you work on it. This keeps the polygon count low and lets you work well with SketchUp. You can then either replace these before you create renderings or—depending on your rendering software—replace them during the rendering process (outside of SketchUp).

TIP

Make sure the components' coordinate axes (the ones that you see in the component editor after you double-click a component) are at the same location—for example, at the center of the bottom of the objects. Otherwise, replacing the components will shift them.

■ **Don't use textures that are too large.** If you use image-based textures, use only the image size that you need. Images can have very high resolutions, and several images with textures that are too large can slow down SketchUp significantly.

Your Turn!

The following tasks allow you to practice the topics in this chapter:

1. **Create Your Own Template**

   To customize SketchUp, start a new file and turn it into your own template by making various modifications (adjust fonts, colors, etc.). Activate it as the default template. Finally, create a new file using this template and model something with it. Add dimensions and text, and print a view to show off your "style."

2. **Build Your Own Trellis**

   Take the examples in this chapter as a guide to model your very own garden trellis. Customize the end cuts of all beams, and add furniture, a few people, and whatever else you see as appropriate. Finally, print several views of the trellis (some perspective views and some orthographic views such as "Front" and "Top"). Add enough dimensions and detail so that a builder could actually construct it from your instructions.

3. **Model Your House**

   Use SketchUp's modeling tool set to model your own house. Approximate the outside dimensions and the roof slopes, and create a model with an average level of detail (don't include details that are too small). If you like, read SketchUp's help files on adding photos to the walls as textures and follow those instructions. You can then upload the model to the 3D Warehouse and get it included in Google Earth's "3D Buildings" layer.

Chapter 3
Using SketchUp to Inform Your Designs

This chapter goes beyond using SketchUp as a simple 3D modeling tool. While this software is excellent at that, 3D modeling in SketchUp can and should be about more than just *representing* your designs—you can use it very effectively to *inform* your designs.

The following sections cover using group- and component-based modeling (for detailing, assembly planning, and other tasks), working with dynamic components, and using geo-based modeling effectively.

Key Terms:

- Component-based modeling
- Reporting component attributes
- Assembly-based modeling
- Trimming and cutting in SketchUp
- Using and making Dynamic Components
- SketchUp and Building Information Modeling (BIM)
- Architectural space programming with SketchUp
- Geo-based modeling

Group- and Component-Based Modeling

As you saw in Chapter 2, using groups and components can clean up your SketchUp model, keep it organized, create 3D geometry efficiently, and permit easy reuse of modeling objects.

You also learned that *groups* in SketchUp are useful for logically grouping individual geometry (e.g., assembling six faces into a brick), while *components* are appropriate if we want to reuse the exact same component multiple times. In addition, components can attach ("glue") themselves automatically to other geometry (any faces, horizontal faces only, etc.), or they can always face the camera (like the person model in SketchUp's default template). You can also reuse components easily through SketchUp's Components window or SketchUp's 3D Warehouse online repository (which can be found at http://sketchup.google.com/3dwarehouse).

This section explores groups and components a bit further.

Effective Use of Groups and Components

Let's look at a mock-up of a stud wall assembly as an example of what we can do when we are consistent in using groups and components to organize our models. The following is a reasonably detailed model of a common part of house construction. This method is very applicable to construction and architecture (because of a building consisting of actual components like bricks and studs), but it should be mentioned that it is of equal value if you use SketchUp for woodworking, interior design, robot building, and many other tasks.

It is possible to model this construction detail in a variety of ways. If you wanted, you could start by drawing the base and use the Push/Pull tool to extrude the footing. You could then draw new outlines on it and extrude the foundation wall. After this, you could make your way upward simply by drawing outlines on surfaces and extruding them. This modeling approach is similar to modeling something out of a block of clay (by pinching, pulling, and pushing). You would finish your SketchUp model by adding appropriate textures to all faces and making the model look like **Figure 3.1**.

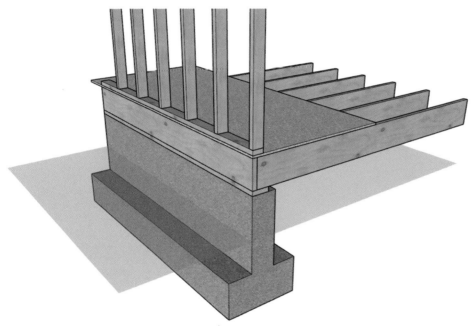

Figure 3.1: A detailed mockup of a construction detail

While this approach visibly gets you the result you want, it is not very flexible and—worst of all—it limits you in what you can do with this model later. As you can see in **Figure 3.2**, simply moving some faces distorts and ruins the entire model (after all, everything is attached if no part of the model has been grouped).

Because our case deals with many individual objects (the concrete foundation, the studs, the joists, etc.), it is much better and more efficient to approach modeling this detail by either grouping each object or turning it into a component while we construct it. In essence, we are creating an assembly in such a way that we could at any point take its components apart—as you can see in the exploded view shown in **Figure 3.3**.

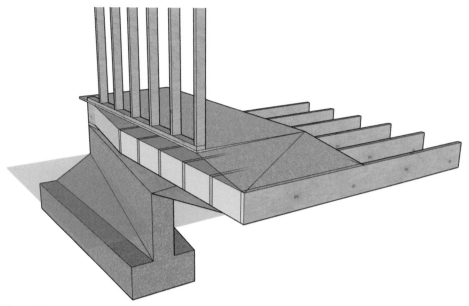

Figure 3.2: Moving parts of an ungrouped model can destroy it

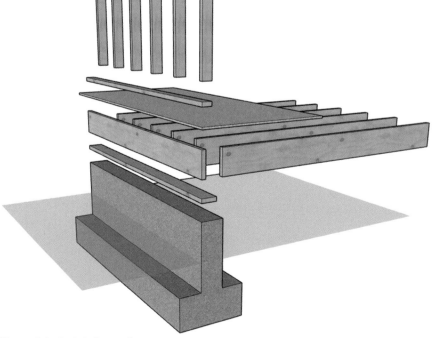

Figure 3.3: Exploded view of assembly made of groups and components

In this detail (**Figure 3.3**), every logical object is separated from the others and is its own entity. What you can't see in this visible representation is which object is a group and which is a component. As discussed before, components work best if you need many similar copies (that are only different in scaling, rotation, and insertion location). Therefore, all of the wood pieces (joists, studs, plates, etc.) are components. Because there is only one copy of the foundation in the model, it was equally efficient to use it as a simple group, although a component would have worked, too.

What we can do now is explore this model in the Outliner window (see **Figure 3.4**). This will give us more information on how this model has been organized. You can find the Outliner in the Window menu. As you will see, it is often a very useful helper to keep around in your workspace.

The Outliner reveals that there is actually a hierarchy of objects in this model: The joist components are part of a group called "Floor" and the studs are part of a group called "Wall." Outside of these two groups are the foundation and the sill. Everything is then contained in a group called "Mockup Detail 1."

This hierarchy was created by *nesting* components, which means that all the joist components have been grouped and the resulting group has been given the name "Floor." As you may already know, renaming an object can easily be done in the Entity Info window. The Entity Info window is another good candidate to always have open in your workspace. It not only allows you to edit names, but also lets you adjust shadow behavior on a per-object basis and modify the layer on which the object resides. (See **Figure 3.5**.)

You can use the Outliner window in a variety of ways. You could, for example, rearrange named objects and place them into groups simply by dragging and dropping them within the list to any position. You can also use this list in conjunction with the Entity Info window to select individual components and hide/unhide them and adjust their locked status. In any case, it is advantageous if you give objects (groups or components) in SketchUp recognizable names so that you can find them easily in the Outliner window.

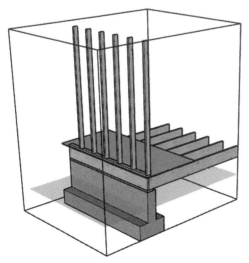

Figure 3.4: Outliner view of frame model

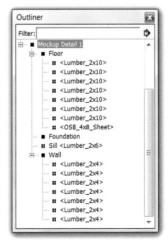

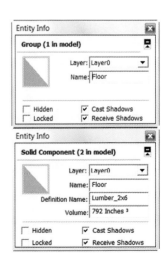

Figure 3.5: Entity Info window for a group (top) and a component (bottom)

TIP

In a large model, the filter function of the Outliner window helps reduce the number of components and focus on what you are looking for. Just make sure you include useful text in your naming (e.g., "building1-floor3-window6") so that you can search easily.

PRO ONLY

Generating Reports of Groups and Components

If you have the Pro version of SketchUp, you can reap another benefit from using groups and components: You can generate a report from your model. To do this, select the objects you want to have reported (the entire mock-up in our case) and go to **File → Generate Report…** This opens the following dialog, where you can specify whether you want the report to be generated as an HTML file (a webpage for viewing in a browser or opening with Word) or a CSV (comma-separated values) file (for import into, for example, Excel). See **Figure 3.6**.

The report generated from our framing model then looks as shown in **Figure 3.7**.

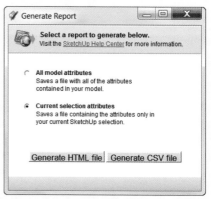

Figure 3.6: Generate Report dialog

PATH	DEFINITION NAME	ENTITY VOLUME	LENX	LENY	LENZ	MATERIAL
Mockup Detail 1	-		137	106.25	145	
Mockup Detail 1/Mockup Detail 1	Lumber_2x6	792	96	1.5	5.5	Wood - SPF
Mockup Detail 1/Foundation	-	46080	24	96	36	Concrete
Mockup Detail 1/Wall	-		3.5	96	97.5	
Mockup Detail 1/Wall/Lumber_2x4	Lumber_2x4	504	96	1.5	3.5	Wood - SPF
Mockup Detail 1/Wall/Lumber_2x4	Lumber_2x4	504	96	1.5	3.5	Wood - SPF
Mockup Detail 1/Wall/Lumber_2x4	Lumber_2x4	504	96	1.5	3.5	Wood - SPF
Mockup Detail 1/Wall/Lumber_2x4	Lumber_2x4	504	96	1.5	3.5	Wood - SPF
Mockup Detail 1/Wall/Lumber_2x4	Lumber_2x4	504	96	1.5	3.5	Wood - SPF
Mockup Detail 1/Wall/Lumber_2x4	Lumber_2x4	504	96	1.5	3.5	Wood - SPF
Mockup Detail 1/Wall/Lumber_2x4	Lumber_2x4	504	96	1.5	3.5	Wood - SPF
Mockup Detail 1/Floor	-		97.5	106.25	10	
Mockup Detail 1/Floor/Lumber_2x10	Lumber_2x10	1332	96	1.5	9.25	Wood - SPF
Mockup Detail 1/Floor/Lumber_2x10	Lumber_2x10	1332	96	1.5	9.25	Wood - SPF
Mockup Detail 1/Floor/Lumber_2x10	Lumber_2x10	1332	96	1.5	9.25	Wood - SPF
Mockup Detail 1/Floor/Lumber_2x10	Lumber_2x10	1332	96	1.5	9.25	Wood - SPF
Mockup Detail 1/Floor/Lumber_2x10	Lumber_2x10	1332	96	1.5	9.25	Wood - SPF
Mockup Detail 1/Floor/Lumber_2x10	Lumber_2x10	1332	96	1.5	9.25	Wood - SPF
Mockup Detail 1/Floor/OSB_4x8_Sheet	OSB_4x8_Sheet	3825	48	106.25	0.75	OSB
Mockup Detail 1/Floor/Lumber_2x10	Lumber_2x10	1332	96	1.5	9.25	Wood - SPF
TOTALS	-	63549	1750	533.25	384	-

Figure 3.7: Model data imported to Excel (some columns removed)

As you can see, not only does this report contain the objects, their hierarchy locations, and component definition names; it also includes their locations in space (not shown in **Figure 3.7**), their volumes, their materials, and their sizes (with LENX, LENY, and LENZ representing lengths in the three main directions—this information is useful when objects have been scaled).

You can now use this data to create various kinds of schedules: door/window schedules, material lists, concrete volume estimation, and more. As you will see, if you make a component "dynamic," then you can add any attribute to it and report it using this method, too.

TIP

Materials will be reported correctly in this table when you apply a properly named material to the component as a whole. Materials included *within* the component (i.e., applied to faces only) are not being reported this way.

TIP

If you don't have SketchUp Pro or you want a little bit more control over how components appear in a cutlist-type component report, look for the Cutlist and Layout plugin in Chapter 4.

Although not specifically mentioned here, adding layers (using the Layers window) gives you an additional method to combine similar objects. This can be useful, for example, if the floor group of components in the preceding situation consisted of existing and new joists and you wanted to be able to separate them visually.

In that case, you would add "Existing" and "New" to the list of available layers in your model and adjust entity properties by putting joists on the appropriate layers. You could then display the existing and remodeled construction situation simply by turning layers on and off.

The example in this section shows how beneficial it can be to assume a rigorous group- or component-based modeling workflow. This becomes even more useful when we include a good naming scheme. At that point, we are starting to use *data* in SketchUp (in addition to plain geometry).

A good real-world example where this can be employed even on very large projects is the case of 4D (3D plus time) or 5D (4D plus cost) construction planning. Using those tools (one example is shown in **Figure 3.8**), construction models can be enriched with timeline data, and the construction process can be visualized (and scheduled), taking into consideration estimated component fabrication times. Commercial software packages that do this with or in SketchUp are 4D Virtual Builder, Envision, and Synchro.

Another example where groups and components can be useful to convey your message is the case of historic reconstruction. As **Figure 3.9** shows, timeline renderings can employ a model, which has been broken into components by construction dates, and then easily display this as a timeline. In this case, you would use a combination of layers and the Outliner window to organize (and show/hide) buildings. This could even be animated using SketchUp's tabbed animation feature and saved as a video (see Example 3.2 for instructions).

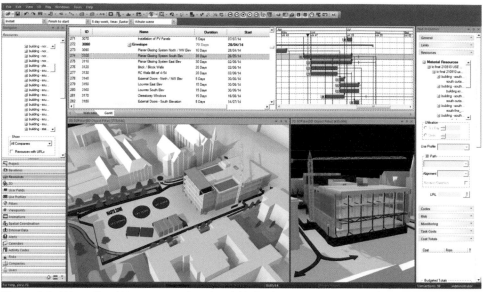

Figure 3.8: Using a SketchUp component-based model in Synchro, a 4D construction scheduling software *(Image used by permission of Synchro Ltd.)*

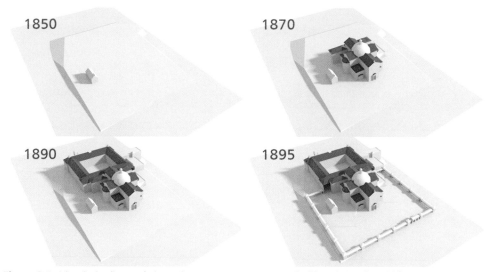

Figure 3.9: Historic timeline renderings using components to separate buildings (original model from 3D Warehouse)

Modeling with Manufacture in Mind

When working with components as shown in the preceding section, it might be tempting to ignore small modeling inaccuracies. This can be the case especially when there are oblique angles or overlaps. To illustrate this, consider two overlapped pieces of wood, as shown in **Figure 3.10**.

Figure 3.10: Overlapping components (left and middle: simple overlap; right: cutout)

Although the left image shows an overlap that looks appropriate, removing the top piece reveals that it simply consists of two components overlapping without any resolution of the intersection. The third image then illustrates how an overlap should be resolved: exactly the way this example would actually be built—with a notch.

Fortunate for us, SketchUp is well equipped to help us model intersections like these. All we need to do is use the Intersect with Model tool (which requires some manual cleanup) or SketchUp Pro's Solid Tools. Once we have this notch modeled in SketchUp, we have not only a proper component-based model but also one that can be built as modeled.

You can see a larger example in **Figure 3.11**, which is an assembly of four overlapping pieces of wood, where each one is angled 4 degrees off the horizontal. As shown in **Figure 3.11**, modeling this and including all proper cuts and notches immediately provides us with a buildable model. All I did to create **Figure 3.12** was copy one component and add dimensions in SketchUp.

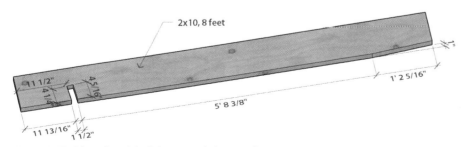

Figure 3.11: Model of four overlapping wood beams

Figure 3.12: Dimensioned single beam, ready for manufacture

So what is the benefit of using this modeling approach? By actually "building" (as in the sense of virtually constructing) our 3D models, we are incorporating something very useful into our design process: immediate design validation and error checking. For example, you can easily see if a cut didn't work in your 3D model and fix it, which—of course—is cheaper than physically building something and having to redo it.

In addition, virtually building assemblies in 3D allows you to conceptualize the construction process, making it much easier—and less error prone—when you actually have to build something in reality (see **Figure 3.13**). Especially when various layers are involved (in architecture this could consist of interior finishes—structure—skin—exterior shading devices), this allows you to think through the process of matching and coordinating all those layers (and all of the subcontractors involved) at an early, conceptual stage.

Independent of your discipline, you can use this approach to test feasibility of construction details on the job site, workflow processes (in plant layout), furniture assembly in the wood shop, modeling for 3D printing, and many more applications. **Figure 3.14** shows how a large construction firm, Turner Construction Company, uses SketchUp in its New York World Trade Center Transportation Hub project to previsualize structural steel assemblies.

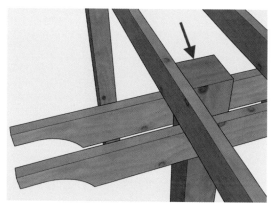

Figure 3.13: Finding an unprotected post top in your model lets you fix it before building it

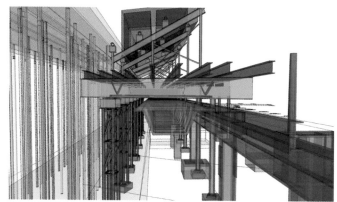

Figure 3.14: Detailed model of box girder steel, precast smoke purge ducts, and temporary shoring *(Image used with permission of Turner Construction Company.)*

Example 3.1: Trimming Groups Using Two Methods

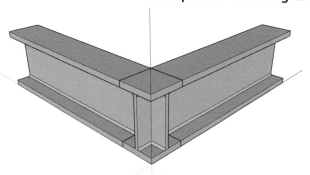

This example uses two methods to trim groups: SketchUp's Intersect with Model tool (which is in both the free and the Pro versions) and SketchUp Pro's Trim feature (from the Solid Tools). The task at hand is mitering two large steel beams at a corner. To start, I downloaded a wide-flange cross-section component from the 3D Warehouse and turned it into two beams, using the Push/Pull tool.

Let's start with the **Intersect with Model** workflow:

1. To line up all necessary geometry, place the two beams in their final positions:

2. If we want to trim one of the beams at 45 degrees, then we need to have a plane that cuts it at exactly that location. This is easily accomplished by drawing diagonals and verticals into the corner to create a face. Extend this face a little bit vertically beyond the beam's edges so that you don't run into any problems when cutting.

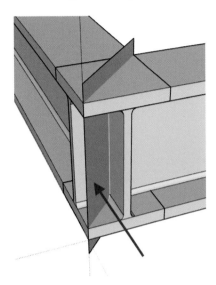

3. SketchUp's Intersect tool creates edges at the interface of objects that are part of the intersection. As you can see in the right-click menu, SketchUp can intersect with the rest of the model, a selection, or the current context. In this case, we will simply use the Intersect with Model feature. It is important to understand that SketchUp will create these intersecting lines in the current context. This means that if you are intersecting the grouped beam and the face from outside the group, all lines will be created outside the group as well. We don't want that here because we need the intersecting lines to "score" our beam sufficiently for us to be able to trim away the overlapping part. Therefore, before you use Intersect with Model, go into editing mode for the beam (double-click it) and select all of its faces. Only then right-click on them and use Intersect with Model.

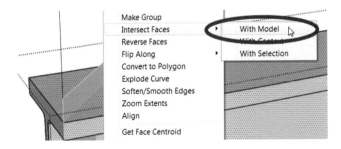

4. This creates scoring lines (#1 in the image shown here). You can now delete everything to the right of those lines (#2 in the same image) to cut away the mitered end.

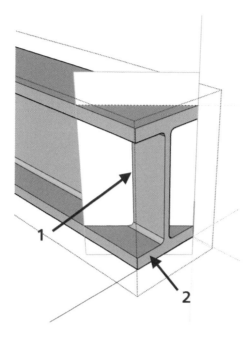

5. At this point, you have a cleanly mitered end of this beam.

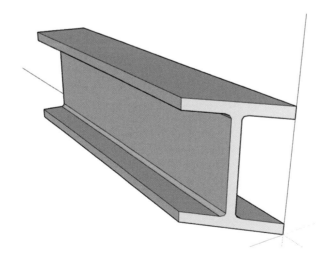

6. Copy and rotate this beam to create the mating part of the corner assembly.

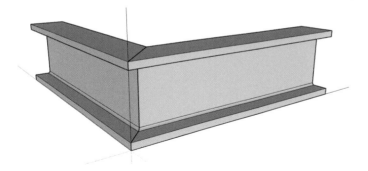

The Trim Tool

If you own SketchUp Pro, then you can accomplish this task much more easily using the Trim tool from the Solid Tools tool set. You can access this tool either from the Solid Tools toolbar or through the **Tools → Solid Tools** menu.

1. Start by creating a diagonal plane in the corner, similar to what we did previously. Because the Trim tool doesn't work with planes but with solids (closed or "watertight" groups and components), we need to extrude the plane using the Push/Pull tool. Do this so that it includes the entire end of the beam you need to cut away. Also, make sure you turn this box into a group.

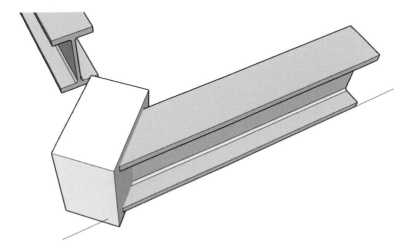

2. You can now start the Trim tool. It will ask you to click two solids—first the box (that you want to cut against) and then the beam (the part you want to cut).

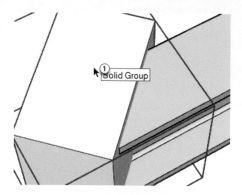

3. This creates the trimmed end of the beam in one step. Just delete the box, and your corner is done.

Example 3.2: Assembly Animation in SketchUp

A SketchUp feature that can be used to great effect with a component-based model is tabbed animation. Whenever you want to save a view in your model, you can add a tabbed "scene" (using the **View → Animation** menu). If you add more of these, you can even get SketchUp to animate through them. It is also possible to export this animation to a video file.

Of great benefit to us here is what is saved with each of these tabs. Not only is the camera location saved, but a tab also remembers which geometry or layers are hidden, where section planes were, which visual style the view had, and what the shadow settings were. For example, you can use these tabs effectively to illustrate shadow movements during a day.

In this example, we will add some scene tabs to the construction frame model from earlier in this chapter to create a small assembly animation.

1. Set up your model with all of the necessary objects. Make sure you group objects and use layers if you need to. See **Figure 3.1** for the model used here.

2. Set up your basic view. This example keeps the view stationary, although you can just as well add different viewpoints in each scene tab. This would create more of a fly-around animation, of course.

3. Now click on the **View → Animation → Add Scene** menu item. This adds the first tab.

4. We will create our assembly animation backward (as a disassembly). Therefore, hide the first object(s) in your model (the wall studs, in my case). Then add another scene (you can do this now by right-clicking on the scene tab).

You can always edit and update a scene by right-clicking on a tab and selecting Update. This updates the scene (and all of its settings) to the current view. If you need more control, right-click on the scene thumbnail in the Scenes window and select from the dialog what you want to update.

5. Hide the next object(s) and create another scene tab. Do this until you are at the end of your disassembly process. You should now see something like this:

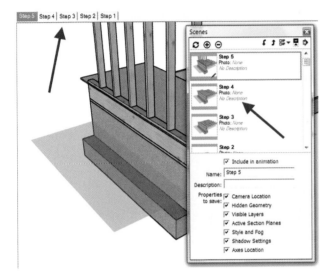

6. The preceding image also displays the Scenes window. This window gives you a good visual overview of the animation tabs and lets you name the tabs. Name them, as shown, in reverse order.

7. Now use the Up and Down arrow keys in the Scene window to reverse the order of the tabs. You should then see a properly organized animation layout like this:

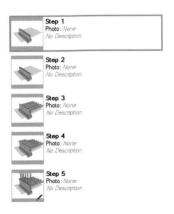

8. Before playing the animation, it is always a good idea to adjust the animation settings (the length tabs are displayed or their transition duration). Do this in SketchUp's Model Info window:

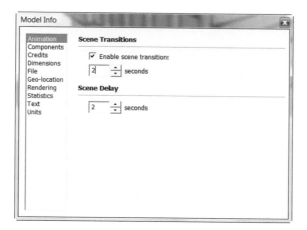

9. You can now view the animation by right-clicking on one of the tabs and selecting Play Animation. A scene-by-scene breakdown for this example looks like this (one scene omitted here):

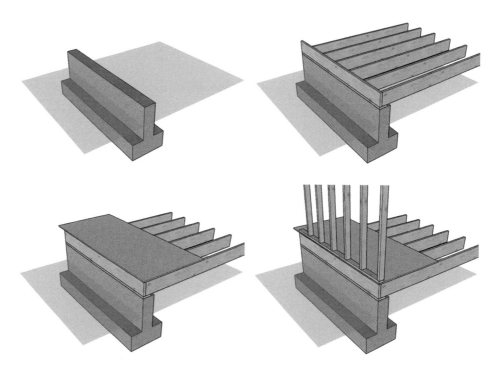

If you want to export this animation as a video, go to **File** → **Export** → **Animation** and save it. To create a good-quality video, shrink SketchUp's window to approximately the final video size (for example, use 1280 × 720 pixels for minimum HDTV resolution) and make sure everything looks good in the scenes before exporting. Also, export the video uncompressed at a high frame rate (25 to 30 frames per second), and edit and compress it later using video-editing software. You can adjust these export options in the video exporter dialog.

Using Dynamic Components to Your Advantage

Dynamic Components were introduced in SketchUp 7 and have in the meantime become a popular addition to SketchUp's core tool set. Their feature set has proven quite useful especially for furniture and building component producers that need to add data to their 3D models, for schedules and other lists, for example.

How are Dynamic Components different from normal SketchUp components? As it turns out, a Dynamic Component's geometry is created exactly the same way a normal "static" component's would be. The dynamic feature gets added later, when the creator decides to add parameters and formulas to the component to embed data or make it responsive to something (e.g., a mouse-click). **Figure 3.15** shows three Dynamic Components—all KraftMaid cabinetry—that were downloaded from the 3D Warehouse.

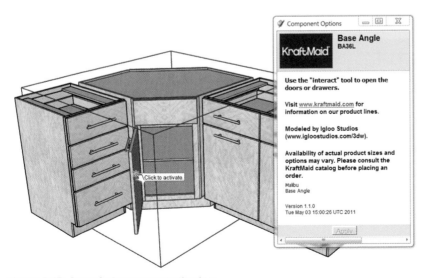

Figure 3.15: Dynamic Components in SketchUp

As you can see, highlighting a component brings up a description containing a part number and a link to the manufacturer's website. Because these are cabinets, it is also possible to click on doors and drawers and watch them swing or glide open. This is useful if you need to check clearances when planning kitchens.

Another way to use Dynamic Components is as parametric objects. **Figure 3.16** shows a bookshelf whose geometry can be changed by editing values such as height, material, and color in a dialog.

Figure 3.16: A parametric (Dynamic Component) bookshelf

Any user interaction with Dynamic Components is done using just a few tools. You can open the Dynamic Components toolbar in the **View → Toolbars** menu to get access to them. As you can see in the graphic, this toolbar contains two or three icons (depending on whether you have the free or the Pro version of SketchUp).

In any version of SketchUp, you will see the hand icon on the left (the Interact tool). This tool permits interaction with a component by clicking, given that the component has that ability. For a good example, click on the shirt of the person in SketchUp's default template ("Susan" in version 8)—it will change color randomly (see **Figure 3.17**).

The tool next to it brings up the Component Options window, which is also in both versions of SketchUp. Whenever you highlight a Dynamic Component (using the standard selection tool), this dialog shows all of the component's properties and attributes. You can see this window in **Figure 3.16**.

Figure 3.17: Interactive Dynamic Component behavior

Component Attributes Window

If you own SketchUp Pro, you will see another button. This one brings up the Component Attributes window (**Figure 3.18**).

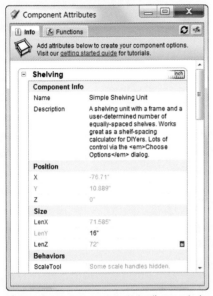

Figure 3.18: A component's Attributes window

In this window you can create attributes and make a "static" component dynamic. You can also always come back to this window and edit any of the parameters or formulas.

Example 3.3: Adding Dynamic Components to Your Model

This example allows you to start working with Dynamic Components by assembling a small kitchen. Feel free to use any manufacturer's cabinets that are in the 3D Warehouse. Just be consistent—otherwise, things might not fit. This example uses SketchUp's DC Cabinets, which you can find here: http://sketchup.google.com/3dwarehouse/cldetails?mid=e89fe26ad1304488de650492e4 5fb14f (short link: http://goo.gl/eK4od).

1. Start by assembling the components you need. For this example, I downloaded two base cabinets and two tall cabinets. Although you can do this from the Component window in SketchUp, too, finding multiple components might be easier by browsing to the website in an external browser, downloading them, and then dragging the component files into

SketchUp. They will then show up immediately at your mouse cursor and you can place them anywhere in your model.

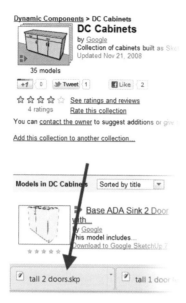

2. Now assemble the kitchen. If you like, add some walls and a floor and any other items such as windows. In the following image, I also added a countertop and some spacer boards at the corner. You could color the cabinets by painting any material on them (I used a wood veneer here). As you can see, the components have been set up so that only the fronts are being painted.

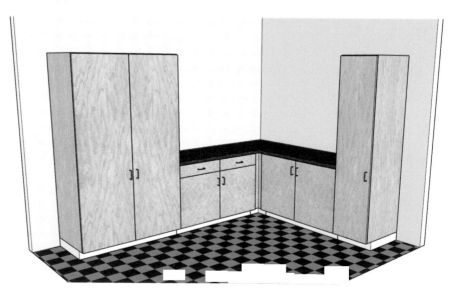

3. Select the edge cabinets and open the Component Options window from the Dynamic Components toolbar. As you can see here, you have several options available for these components. Don't forget to make sure the Finished Ends property is set appropriately. This adds color to the sides of the cabinets as well.

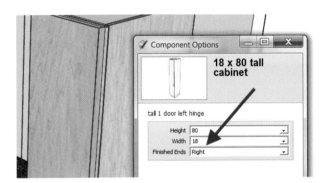

4. You can now use the Inspector tool to your advantage. Click on some drawers and doors. As you can see, they will slide out and swing open to their fully extended positions. This can give you some feedback on whether there are narrow spots and overlaps (especially in the corner). Because I added a filler strip in the corner (instead of just abutting the cabinets), everything fits well here.

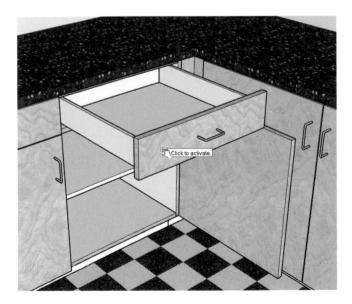

This short example gives you a quick overview of how to use Dynamic Components in your designs. Feel free to browse some other Dynamic Component examples in the 3D Warehouse to see how they can be used for landscape design, manufacturing, construction, and other applications.

Next, let's look at how we can make our own Dynamic Components. Because there are quite a few functions available for these, learning how to make them can become quite involved. Fortunately, the makers of SketchUp produced several sample components that convey this in more depth than can be covered here. You can find them at http://sketchup.google.com/3dwarehouse/cldetails?mid=d86b819368a7afe98095ba2080357053 (short link: http://goo.gl/vZt2Z). Also, please note that the ability to create Dynamic Components is included only in SketchUp Pro.

PRO ONLY

Making a Dynamic Component

In this example, we will create a 2x4 stud wall as a Dynamic Component and give it the capability of reporting material counts.

1. To start the wall, we need to begin with a basic amount of geometry. Draw a rectangle on the ground plane that measures 4" by 4'. This creates the base of the wall component. Because this is intended to be an interior wall, its width is 4" (3½" for the studs plus a layer of 1/2" GWB Sheetrock on each side).

2. Extrude this rectangle to the total wall height of 8'3" (assuming a single top and bottom plate).

3. Use SketchUp's standard Make Component tool to turn this box into a component. In the Create Component dialog that opens, give it a name and select Glue to Horizontal (because we always want this to be a vertical wall).

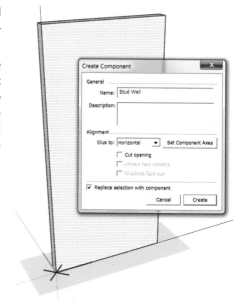

4. At this point, the wall is a "static" component that you could use like any other in SketchUp. We need to turn it into a Dynamic Component, however. Therefore, open the Component Attributes window using the button on the Dynamic Components toolbar. Your workspace should look like this (when the wall component is selected):

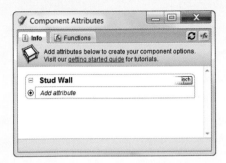

5. As you can see, this dialog contains an Info tab that shows you all of the already defined attributes. It also has a Functions tab, which gives you an overview of all the functions that you can use with Dynamic Components. Let's add some attributes now. Click on Add Attribute to bring up a selection of common attributes. To begin, we will add the LenX and LenZ attributes.

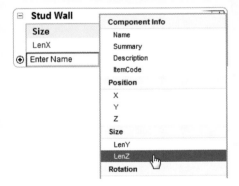

6. These attributes will report the lengths along the component's (internal) red and blue axes. At this point, they should show values of 48" and 99", respectively.

7. Let's add a calculated attribute now. Click on Add Attribute and type in a name (instead of picking one). Name it GWB_Area (you can't have spaces in names, hence, the underscore).

8. This attribute calculates the area of Sheetrock on both sides of the wall. Therefore, the formula for this attribute must be `=LenX*LenZ*2`. Enter this in the box to the right of `GWB_Area`.

9. The `GWB_Area` field should now report a value of 9504 (square inches of Sheetrock area). You can click on the button that reads "=fx" in the top-right corner of the dialog to toggle between value view and function view.

10. Add another calculated value; this one reports the number of 4 × 8 GWB sheets that are needed. Call it `GWB_Sheets` and give it the formula `=GWB_Area/4608`.

11. Now click on the arrow to the right of this formula to bring up the attribute's options. Make sure to select "User can see this attribute," give it a readable name, and display it as a decimal number.

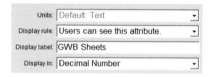

12. Finally, add another attribute, `Studs_Num` to the list. Give it the formula `=ROUND(LenX/16+1)`. As you may have guessed, this attribute reports the (rounded) number of studs for this section (assuming 16" on-center spacing). We also used one of SketchUp's built-in functions for this attribute ("ROUND"). Don't forget to change the attribute's options so that the user can see this one, too.

13. To finish this component, let's add one more built-in attribute from the drop-down list: `ScaleTool`. This attribute lets you define which scale handles are shown to the user. For our wall component, we only want the user to be able to stretch the wall using the "Scale along red (X)" handle. Uncheck everything else.

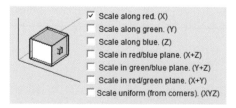

14. The finished Dynamic Component should look like the following image when you use the scale tool. All you should see is the scale handle in the red direction (the X axis).

You can now stretch this component to your liking and see the reported attributes update in the Component Attributes and the Component Options windows. If you distributed this Dynamic Component to SketchUp Free users, they would be able to place it, stretch it, and read the values from the Component Options window. They would not be able to edit it, though.

After inserting several wall component instances into your model, it is possible to use the **Generate Report** menu item to report stud number and Sheetrock sheet requirement for the entire model (or a selection).

As a reference, you can find a complete list of available Dynamic Component functions and predefined attributes in Appendix E.

If you don't have SketchUp Pro, don't worry. You can still go to this book's companion website and download this wall component for use in your models. You just can't edit its definition.

Where Does SketchUp Fit into the BIM Workflow?

SketchUp (free and Pro) has been used for years by many architecture, engineering, and construction (AEC) and related professionals to solve various tasks over the course of many building projects. Some firms use it in the early, conceptual phase to study design options and maybe perform energy analysis. In later design phases, SketchUp has also found a wide following as a rendering preprocessing tool. Geometry can easily be imported from other software and various low-cost rendering software packages can be used to create highly professional renderings (see Chapter 5 for more on this). In addition—as you saw before—SketchUp can be used successfully in construction planning for assembly mock-ups, 4D scheduling, and site layout design.

On the other hand, a firm might decide to use SketchUp Pro as its sole CAD tool and benefit from LayOut's ability to create production-ready construction documentation. LayOut is a drafting add-on that ships with SketchUp Pro.

With the recent surge in demand for virtual construction tools and Building Information Modeling (BIM) emerging as both a tool and a process in digital construction, the question validly arises as to where SketchUp fits into all of this. After all, BIM is all about 3D modeling and data integration. And as you know by now, not only is SketchUp capable of 3D modeling, it also features ways to include data in your models.

While SketchUp, at its heart, is *general* 3D modeling software (and not tied to a particular industry), it can offer these avenues for use in architectural, engineering, or construction planning:

- **As a massing-modeling tool**—You can easily do massing-modeling in SketchUp by importing existing 3D buildings from the 3D Warehouse and then blocking out new buildings using SketchUp's tool set. Many larger BIM applications (e.g., Autodesk Revit and Graphisoft ArchiCAD) are able to import SketchUp models and use them to generate more detailed building models. **Figure 3.19** illustrates this for Revit.

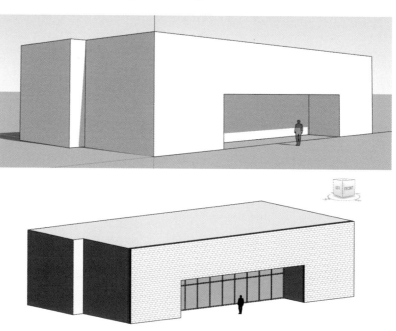

Figure 3.19: Importing a SketchUp model into Revit (front walls were created using the Wall by Face tool)

- **As a programming tool**—When laying out building programs, it is possible to easily "block" out areas and spaces and arrange them in SketchUp (see **Figure 3.20**). It is even possible to create these spaces based on data from external software—examples are the Onuma Planning System and Trelligence Affinity.

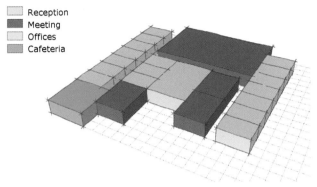

Figure 3.20: Space planning in SketchUp

- **As a building energy analysis tool**—With a building's energy performance more frequently being included into early stages of design, it is possible to use SketchUp as a rough modeling tool (these models often need to consist only of simple faces to work well) and an analysis preprocessor. Some of the available software then exchanges the SketchUp-created energy model with an external energy analysis program or with other BIM applications using the gbXML (green building XML) format.

A list of currently available plugins and an example are included in Chapter 4.

- **As a mock-up building and constructability analysis tool**—While BIM software is able to create 3D models of sometimes very complex buildings, these are often not executed at a level of detail that might include individual studs or bricks. If 3D assembly models are required to evaluate or explain details, it is sometimes more efficient to do them in SketchUp and later import them into the construction documentation set.

As shown in the preceding examples, SketchUp can be used very effectively to create mock-ups and test assemblies as long as a strict component-based modeling approach is followed.

- **As a geo-based modeling tool**—One of SketchUp's strengths is that it is tightly integrated with geo-based tools, which are:

  - The **ortho-photographs** (mostly satellite photography) that you know from Google Maps and Google Earth. You can add them to your model and thereby geo-locate your model. You can even export your model to Google Earth afterward.

  - **Terrain models** (at varying detail levels) for the entire globe. A terrain is automatically imported when you geo-locate your model using the tool shown in **Figure 3.21**. Use the Toggle Terrain button to show it.

    Chapter 4 includes an example using SketchUp's terrain information to create a layered terrain model.

  - **Street View imagery.** These are car-top-height panoramic images of almost every accessible street in many cities. You can include them in your model using the Photo Textures window (for example, to texture the sides of a building).

- **Oblique aerial imagery.** While mostly available only for cities and major urban areas, these images provide angled views of buildings showing all facades. At this point, oblique imagery is accessible only using Google's Building Maker tool, but building models created with it can be reimported into SketchUp.

You can access Building Maker here: http://sketchup.google.com/3dwh/buildingmaker.html.

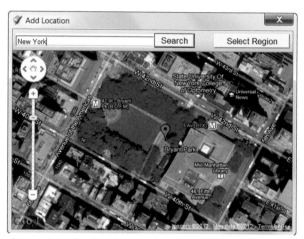

Figure 3.21: Geo-locating a model using the Add Location tool

You can exchange data and building models from SketchUp with other software in a variety of ways. The free version of SketchUp allows you to export 2D images, 3D models (as COLLADA DAE or Google Earth KMZ files), and animations. The Pro version adds to that AutoCAD DWG and DXF export as well as some generic 3D formats like 3DS, FBX, OBJ, XSI, and VRML.

To import 3D models into SketchUp, you can use either 3DS, DEM, DAE, or KMZ file formats in the free version. The Pro version adds to this AutoCAD DWG and DXF import.

Beyond these formats, many 3D software packages now include direct SKP file import, which makes it even easier to use a SketchUp model in another program. If this is not available, then a good file format, which is universally accepted and supported by all versions of SketchUp, is the COLLADA format. This is also the format Google Earth uses (in its KMZ files) and the one in which models in the 3D Warehouse are stored (besides the original SketchUp file format).

Example 3.4: Program Planning with SketchUp's Dynamic Components

When laying out a space program for a building, a common approach is to use rectangular "blocks" to represent spaces (offices, auditoriums, etc.) and then arrange them visually to outline the total space requirement and relationships among them. In this example, we will use simple rectangular Dynamic Components to help us with this task.

If you don't have SketchUp Pro, you can still do this example. Simply download the Space component from the companion website and insert it into a model. Skip to step 7 to start using this component.

Making a Space Dynamic Component

1. Create a rectangle on the ground plane. At this point, neither size nor location is important. Turn the rectangle into a component and call it "Space."

2. Add all of the attributes as shown in **Figure 3.22**.

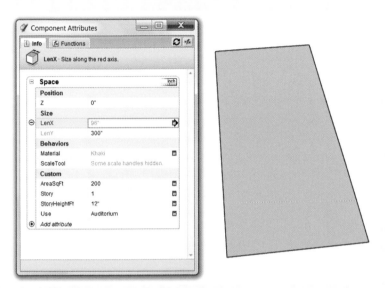

Figure 3.22: Attributes for Space Dynamic Component

3. Add these formulas: for Z use `=(Story-1)*StoryHeightFt*12` and for `LenY` use `=AreaSqFt/LenX*144`. This determines the elevation of the component based on the story on which it is located. In addition, component size is always calculated based on the square-foot area provided by the user. This is helpful when the user stretches the component.

4. For the `Material` attribute, add several of SketchUp's color names that are to your liking. Users are able to color-code spaces this way. Alternatively, you could program into the component the ability to automatically select a color based on the chosen use.

5. In the settings for the `ScaleTool` attribute, deselect everything except "Scale along red (X)" and "Scale uniform (from corners) (XYZ)."

6. Set the visibility of all attributes that show a dialog icon on the right in **Figure 3.22** so that the user can see and edit them.

7. Open the Component Options dialog. It should show all of the accessible attributes such as color, use, story height, story number, and square footage (see **Figure 3.23**).

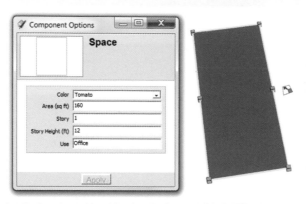

Figure 3.23: Options for the Space Dynamic Component

8. When you start the Scale tool, you will be able to see the handles that are available. Use them to stretch the space component in various directions. As you can see, its area is automatically limited by the Dynamic Component to the entered value: 160 square feet, in this case. (See **Figure 3.24.**)

Figure 3.24: Scaling the Space component

9. To roughly lay out a building, I used SketchUp's Grid plugin to add several grids (with a spacing of 4′) at the story heights. You can download it here: **www.sketchup.com/intl/en/download/rubyscripts.html**.

Having a grid is not absolutely necessary; however, it provides a good reference for snapping.

10. After you do this, add several of your newly created Space components from the Component window into the model, add use and size data in the Component Options window, and place them wherever you like in your model. You can very quickly and efficiently lay out a building program based on this. (See **Figure 3.25.**)

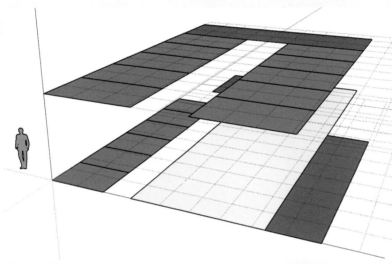

Figure 3.25: Space planning using our component

There is a lot you can do with this seemingly simple approach. Here are some ideas:

- Use the Generate Report tool to *create schedules* of these spaces and export them to Microsoft Excel. If you need to attach more data to each space, do this in the Component Attributes window, as previously shown.
- Learn Ruby scripting (covered in Chapter 6) and *create spaces automatically based on CSV data* provided from another application (e.g., Microsoft Excel).
- Export all of the spaces as a 3D DWG/DXF (or any of the other available file types) and *import them into BIM software* to start adding walls and other features.
- *Export your space model to Google Earth* and view it in its urban context.
- Upload your SketchUp model to the 3D Warehouse and use the Warehouse's versioning abilities to *share and collaborate on programming.*

Geo-Based Modeling

Whenever you work on a project that is physically located anywhere on earth, you might benefit from SketchUp's abilities to give your model a global location, too. As you saw in **Figure 3.21**, assigning a location to a model is very easy with SketchUp's Add Location tool.

Once you have your model accurately geo-referenced, you are able to use this information for anything from adding terrain data to shading analysis. Both of these can inform your designs in a variety of ways. **Figure 3.26** shows an example of this approach.

In **Figure 3.26**, it was not necessary to model an entire building. Of interest here is interior light distribution for various window shading devices. Having modeled the single office in a geo-referenced SketchUp model gave enough information to compare them. As you can see in **Figure 3.26**, the office was simply created on top of the snapshot image that came into SketchUp through the Add Location tool.

Another example is the design of a gardener's clock (basically a sundial), as shown in **Figure 3.27**.

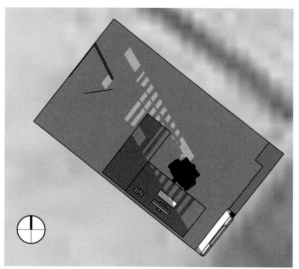

Figure 3.26: Using location information for accurate shading studies for an office

Figure 3.27: A gardener's clock, designed using SketchUp's shadows feature

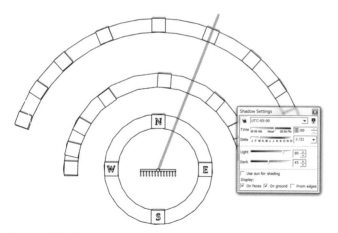

Figure 3.28: Plan view of the sundial with the Shadow Settings window

As you can see in **Figure 3.28**, the placement of the square stones was determined by adding a location to the SketchUp model and then setting SketchUp's shadows to the accurate time and date. The inner circle represents the spring equinox (3/21), and the outer ring is accurate at the summer solstice (6/21). Changing time to each full hour on those dates produced the locations of the hourly stone markers.

As mentioned, you can use geo-located SketchUp models in Google Earth and many other geo-based software packages (e.g., ESRI ArcGIS).

One additional aspect of geo-based modeling that is not covered in this book is phototextured building modeling. To learn more about this, I encourage you to read Aidan Chopra's book *Google SketchUp 8 for Dummies* or visit SketchUp's tutorial site: http://sketchup.google.com/intl/en/yourworldin3d/index.html.

Your Turn!

The following tasks allow you to practice the topics in this chapter:

1. **Create an Exploded View of an Object**
   Similar to **Figure 3.3**, create an exploded view of an assembly or an object in which all the parts are spaced slightly apart. This is a great way to explain how this object was produced.

2. **Create an Assembly Animation**
   Take any assembly or process you are interested in (e.g., a construction detail or a factory layout) and create an assembly animation. Use the technique from Example 3.2 to accomplish this. Also export your animation as a video (and share it with the world by uploading to a video-sharing website). If you want to physically animate objects in SketchUp, look in Chapter 4 for some ideas.

3. **Create a Wooden Puzzle Toy**
   In SketchUp, design a realistic 3D puzzle toy, which is produced using overlapping wooden blocks and cutting them at intersections. The toy must consist of more than three parts and needs to be able to be disassembled and reassembled. If you like, produce it physically afterward in the wood shop to see how well executed your design was.

4. **Make a Dynamic Component**
   If you have access to SketchUp Pro, make a Dynamic Component that is relevant to your discipline. Add it to a model in SketchUp and then report its parameters in an Excel spreadsheet.

5. **Create a Space Plan**
   Use a fictitious building space program (several offices, meeting rooms, assembly areas, etc.) or a real building's space layout and create a 3D space plan using the technique shown in Example 3.4.

6. **Plan a Local Park**
   Use SketchUp's location tools to locate your model and get satellite imagery as well as terrain data. Plan a small park (or assembly area) and use location-based features for your design, which might include the terrain or shading effects.

Chapter 4
Using Plugins Effectively

In this chapter, we'll explore the many plugins that are available to extend SketchUp and use them for tasks that can't be done using SketchUp alone. Those plugins are often freely available, they are easy to install, they add a wide range of functionality, and you can use as many or as few as you need for a given task.

Some of the things that are possible with plugins are physics simulation, advanced creation/deformation of objects, organic modeling, fabrication preprocessing, building energy analysis, and—as we will see in Chapter 5—photorealistic rendering.

Key Topics:

- Where to find plugins and how to install them
- Plugins for general 3D modeling
- Plugins for architectural modeling
- Plugins for digital fabrication
- Plugins for data integration
- Plugins for animation
- Plugins for analysis

What Does a Plugin Do?

Arguably one of the most forward-looking and best features of SketchUp (and one that was implemented early in the software development) was the inclusion of a scripting language called Ruby. Having many functions available in this scripting language effectively made it possible to control every aspect in SketchUp using only a few lines of code. It was thereafter possible for any user to extend the functionality of SketchUp by writing some Ruby code, posting it on the Web, and letting other people use the tool they created, too. Any available plugin can be used with the free version of SketchUp, which is a tremendous benefit for users on a budget who want to get access to powerful 3D CAD software.

Over the past years, this has expanded from small pieces of code that come in one file and add one tool to SketchUp to sometimes very complex software packages that provide an entire tool set and a visual user interface and that allow SketchUp to exchange data with external software such as a photorealistic rendering or a building energy analysis program.

The number of available plugins has increased, too. One of the oldest plugin repositories, Didier Bur's Ruby Library Depot, at the time of this writing was listing over 930 plugins—not even counting many commercial plugins that are available on other websites.

This chapter presents some of the more common plugins. Due to the large number of available plugins (and the small space available here), you are encouraged to keep browsing plugin repositories and following new plugin announcements—especially when you run into a problem that SketchUp alone can't help you solve.

Getting Plugins

Before you can use any plugins in SketchUp, you obviously have to acquire them. While SketchUp made it very easy to find (and share) user-generated 3D SketchUp models by providing the 3D Warehouse, they didn't do the same for plugins. At the time of this writing, there is not one central repository (or even an all-inclusive index) available. There are, however, several community-based websites that feature very extensive plugin indexes and repositories. Each one of them has benefits and drawbacks, and finding the appropriate plugin might require you to browse several of these sites.

Use the following collection of links as your starting point:

- **SketchUp's Plugins Page [1]**—A page on the official SketchUp website that lists some (mostly commercial) plugins. www.sketchup.com/intl/en/download/plugins.html

- **SketchUp's Ruby Scripts Page [2]**—This page provides an introduction to Ruby scripts (which are essentially plugins) and lists some of SketchUp's demonstration scripts. www.sketchup.com/intl/en/download/rubyscripts.html

- **Ruby Library Depot [3]**—This site has been collecting and indexing plugins for a long time now and might at this point have the largest index of freely available community-supplied plugins. http://rhin.crai.archi.fr/rld

- **Smustard [4]**—A site that provides many free or low-cost plugins for daily tasks. www.smustard.com

- **SketchUcation Plugins Forum [5]**—This forum is a venue for plugin authors to publish and discuss their plugins. You can find many community-developed plugins here. Register to be able to download plugins. http://forums.sketchucation.com/viewforum.php?f=323

- **SketchUcation Plugins Index [6]**—This topic in the same forum features a list of all submitted plugins. http://forums.sketchucation.com/viewtopic.php?p=250026

- **SketchUp Plugin Index [7]**—This site provides an index for commercial and freely available plugins. Plugins can be rated and discussed and then downloaded and directly installed using a Plugin Browser Plugin. http://sketchupplugins.com

- **Extend SketchUp [8]**—A site similar to the preceding one with a focus on community-developed plugins. You can also rate your favorite plugins. www.extendsketchup.com

- **SketchUp Plugin Reviews [9]**—This blog publishes reviews of commercial and community-developed SketchUp plugins. http://sketchuppluginreviews.com

- **Jim's SketchUp [Plugins] Blog [10]**—A blog on SketchUp plugins. http://sketchuptips.blogspot.com

Please note: The numbers in square brackets in the preceding list serve as references as we examine some plugins in the next few sections.

> To make life easier for you, I created a Custom Google Search Engine that searches all of the sites mentioned here. If you would like to use it, follow this URL:
>
> **http://goo.gl/gbO2X**

Installing Plugins

When you search for plugins, you will soon find out that they come in one of four file types: RB, ZIP, RBZ, or an installable EXE or DMG file. There is even a fifth one, RBS, an encrypted version of an RB-file. All of these require different installation methods, which I will explain subsequently. For now, it is important that you know *where* SketchUp plugins get installed.

Depending on your operating system, there is a default location in the SketchUp installation—the Plugins folder, which serves as a home for plugins. If you accepted SketchUp's suggested locations on your hard disk during installation, then you can locate this folder as follows.

On Windows computers:

C:\Program Files\Google\Google SketchUp 8\Plugins\

On Macs:

/Library/Application Support/Google SketchUp 8/SketchUp/Plugins/

(If you are running a SketchUp version other than 8, replace the number in the path with your version, of course.)

Whenever you install a plugin in SketchUp (using any of the methods described here), the plugin files are copied into this folder. Every time SketchUp starts, it looks in this folder and automatically loads all of the plugins that it finds.

TIP

If you use a lot of plugins, then the Plugins folder can become quite crammed. You can make space there (and speed up SketchUp's start-up time) by removing some of the plugin files and saving them somewhere else. You can always return them when you need them again.

Let's look at how to install plugins, depending on the file type.

- **RB or RBS file**—This is simply a Ruby code text file that by itself will contain all the necessary instructions for the plugin. Take this file and copy it into the Plugins folder. Then restart SketchUp and you can use the plugin. If you have an RBS file, then this is just an encrypted version of the plugin. While you can't read its contents, SketchUp can. Install it exactly like you would an RB file.

- **ZIP file**—A ZIP (compressed) file contains not only the RB file(s) but often other necessary files (e.g., images) and folders. Unzip this file using any compression software you like and paste its content exactly as it is into the Plugins folder (this is sometimes called *extracting* the file contents). Important: Make sure you keep its internal folder structure intact. Usually your operating system will open a ZIP file if you just double-click it. If that doesn't work, look for compression software such as 7-ZIP (www.7-zip.org) to help you out. *Please note:* You can also install a ZIP file plugin using the method described for the RBZ file.

- **RBZ file**—This is a new file format that was introduced in SketchUp's Version 8 Maintenance Release 2. If you are running an older version of SketchUp, consider upgrading (it's worth it!) or look for a ZIP version of the plugin. If you have the latest version of SketchUp, then all you need to do is go to the Extensions tab in the Preferences window (see **Figure 4.1**). There you will find a small button labeled "Install Extension. . .." Click that button, browse to the RBZ file, and okay any confirmation dialogs that come up. SketchUp will install this plugin for you, and you don't even have to browse to the Plugins folder. As mentioned before, this method also works with ZIP files.

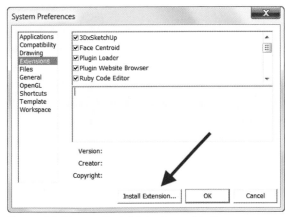

Figure 4.1: The new "Install Extension . . ." button.

In Windows, you should be able to simply copy the URL of an RBZ or ZIP plugin and paste it into the dialog that opens when you click the "Install Extension . . ." button. Windows will download the file for you, and then SketchUp will install it in one step.

- **An installable EXE or DMG file**—If a plugin comes as this type of file, the plugin author will have packaged it in a system-specific installation or setup file (an EXE file in Windows and a DMG file for use on Macs). You can then install it as you would any other software on your computer. Usually this format is used with larger, commercial plugins. Most of these do not require you to do anything other than click a few Next buttons. You might be required to confirm SketchUp's installation location, though. Modify this only if you installed SketchUp in a custom location. Restart SketchUp to activate your plugin.

After installing your plugin, you might be required to restart SketchUp. At that point, any new functionality will be available. Consult your plugin's documentation (if any) to see how you can use it now. Typically, plugins add a menu item to one of the main menus (often under the **Plugins** menu). You might also find new tools in the right-click menu.

It is also possible that the plugin added an entire toolbar. If you can't see the new toolbar, go to the **View** → **Toolbars** menu and make sure you enable it (there should be a check mark next to the plugin's toolbar).

At this point, you are ready to work with your new plugin.

TIP

If you prefer *not* to have all of your plugins load when SketchUp starts and, rather, want to load them "on demand," then use this method:

1. Install my Plugin Loader Plugin from this link: **www.alexschreyer.net/projects/plugin-loader-for-sketchup**.

2. Save your plugins wherever you like—you can even create a folder on a USB drive and save them there. Just make sure you save each plugin with its folder structure intact (exactly as it would be in SketchUp's Plugins folder).

3. Using the **Plugin Loader** menu under the **Plugins** menu, browse to the main RB file of the plugin you want to load, and select it. After loading, your plugin should be available in SketchUp.

Keep in mind that once you restart SketchUp, the plugins you load with this method are no longer available.

Uninstalling and Organizing Plugins

What can you do if you no longer want a particular plugin in SketchUp? As you've learned now, once a plugin is installed in SketchUp's Plugins folder, it is loaded automatically every time SketchUp starts. To uninstall or disable it, you have one or two options (depending on the individual plugin).

If the plugin takes advantage of SketchUp's built-in Extension management system, then it is very easy to disable it and prevent it from loading. You do this in the Extensions tab of the Preferences dialog. As shown in **Figure 4.2**, just uncheck the box next to the plugin's name, and it will not be loaded in the future. If you need it again, just check the box again and restart SketchUp (if necessary).

If you can't find the plugin you are looking for in that list, then your only option is to remove its file(s) from the Plugins folder. This is a little more involved than the previous method because you now have to delete or move the appropriate files from a folder that is buried reasonably deep in your operating system.

To do this, browse using Windows Explorer or Mac's Finder to SketchUp's Plugins folder (as mentioned earlier). If you know which files were installed by the plugin you intend to remove, then simply select all of them. If you don't, some detective work might be required. Remember: Plugin files end in RB or RBS. Also, some plugin authors add their initials at the beginning of the file. In addition, there is often an associated folder that contains more files that this plugin uses (it is not necessary to remove this folder, though).

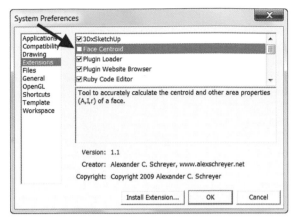

Figure 4.2: Disabling a plugin in the Extensions tab

Once you have found the appropriate files, it might be a good idea to move them to a separate folder (you could create a folder called "Inactive Plugins" for exactly this purpose). The benefit of moving the files instead of deleting them is that if something goes wrong (or you want to have your plugin back), you can always troubleshoot and move files back. You also don't have to search for and download the plugin again in the future.

TIP

If your plugin came as an installable EXE or DMG file, look in your operating system for the uninstall function and uninstall it this way.

Another related issue is the messiness of the Plugins menu in SketchUp once you have installed several plugins. Unfortunately, SketchUp does not give you a way to organize this menu (it simply loads plugins in alphanumeric order based on filenames). If you want to do any spring cleaning here, then a plugin by Smustard [4], the aptly named Organizer, might be able to help you out.

Plugin Overview

The following sections present a selection of the most commonly used and popular plugins. This list is by no means a complete reference. As I mentioned before, there are currently hundreds of available plugins. Use these for the most common tasks but feel free to search the sites mentioned previously if you have a special application that is not covered here.

Unfortunately, the standard for documentation for each of these plugins varies widely. Commercial plugins and those by the more prolific creators (such as Fredo6 and TIG) are usually well documented. For others, read the accompanying text on webpages where you downloaded them.

Whenever I mention a plugin, the name (or alias) is that of the author. I also mention the cost of a plugin (if known) and where you can find it. The number in square brackets refers to the list of website(s) that appeared earlier in this chapter.

A final note before you get going with plugins in SketchUp: Plugin development and especially maintenance are quite involved tasks. Although many plugins have been provided for free to the community, consider donating to the authors if they ask for it and you find their plugins useful. If you get inspired by what is possible in SketchUp and take to Ruby scripting yourself (especially after reading Chapter 6 in this book), then an alternative idea to "pay back" to the community might be to share your own code as plugins.

Plugins for General Modeling

BezierSpline
By Fredo6—Free [5, 6]

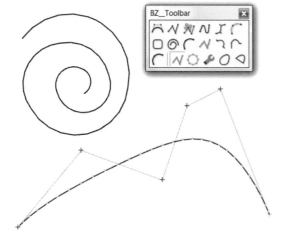

If you need a smooth curve (a spline) for a walkway or a kidney-shaped table, for example, use this plugin to create curves that are determined by control points. You can edit created curves by moving control points and applying various parameters, such as the precision of a curve.

As always in SketchUp, once a curve closes, a face is created. And while edges appear as curves, they do consist of connected polyline segments just like any other curved shape in SketchUp.

BoolTools
By Whaat—$10 [4]

If you don't have the Pro version of SketchUp and are therefore lacking the Solid Tools, then this plugin offers a cost-efficient way to use Boolean operations such as subtract, add, or difference on components.

Components onto Faces
By C. Fullmer—Free [5, 6]

This plugin copies components orthogonally onto all selected faces. The result is similar to what can be seen in the "Placing Components on Faces" example in Chapter 6.

Curviloft
By Fredo6—Free [5, 6]

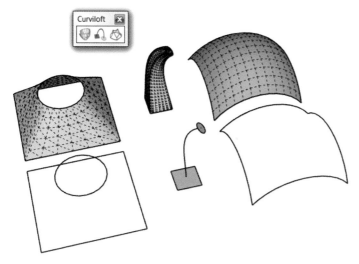

This plugin makes it easy to fit a surface (consisting of quadrilateral or triangular faces) between a set of controlling lines or curves. You can create either a lofted surface, a swept surface, or a surface bordered by edge curves.

Once you pick your controlling surfaces, the plugin gives you many options for editing them before finally committing.

Driving Dimensions
By Menhirs NV—Free/EUR 145 [www.drivingdimensions.com]

This commercial plugin allows you to parametrize your SketchUp model. Adding dimensions using this plugin provides you with a dimensional constraint in your model. You can then modify these dimensions by updating their value rather than by moving objects or stretching them.

Extrusion Tools
By TIG—Free [5, 6]

This is a collection of several formerly published extrusion tools that install using a single ZIP file and are contained in a single toolbar.

The individual tools allow extrusion of lines into faces, edges by edges (one is the shape and the other is the rail along which it extrudes), lofting, multiedge surface creation, a follow-me-type extrusion (e.g., for handrails), and many others.

As you can see in the image shown here, this tool is useful for creating curtains, piping, handrails, lofted shapes, edge-ruled shapes, and much more.

FredoScale
By Fredo6—Free [5, 6]

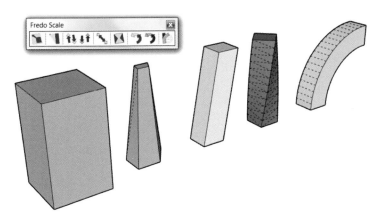

This plugin picks up where SketchUp's Scale tool leaves off. You can also scale with this tool, but more options are available (specifically, *about* which point you want to use each tool). In addition, you can use it to shear, stretch, twist, rotate, and bend any geometry.

In the image shown here, the original shape was a square extrusion. Hidden lines are shown to illustrate how the shape's surfaces were modified.

Geometry Plugins

By Regular Polygon—Free [www.regularpolygon.org/plugins]

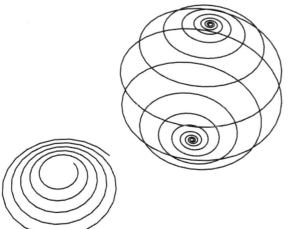

This collection of several plugins allows you to create various geometric shapes that wouldn't be possible using SketchUp. These shapes are Ellipse, Superellipse, Superellipsoid, Sphere, Torus, Loxodrome (Spherical Helix), and Spherical Spiral (Archimedes Spiral).

Another useful geometric tool is **DrawHelix** by J. Folz. It is also free, and you can download it from [10].

Grow

By TIG—Free [5, 6]

This plugin lets you make multiple copies in one step based on equal x, y, and z spacing, rotation about any axis, and scaling.

Joint Push Pull

By Fredo6—Free [5, 6]

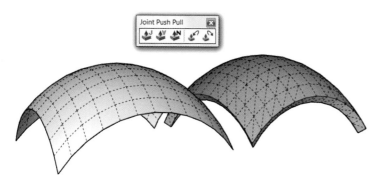

This plugin is an excellent tool for "thickening" surfaces. It allows you to use a Push/Pull-like tool on the connected faces that make up a surface, yet keep everything joined during this process. Available functions include Joint Push Pull, Vector Push Pull (thicken all surfaces in the same direction), and Normal Push Pull.

Main uses for this plugin are (1) adding material thickness to surfaces (e.g., to create a concrete shell), (2) offsetting surfaces, and (3) adding thickness to a surface object to apply a subsurface-scattering material to it in your rendering software.

Lines2Tubes

By D. Bur—Free [3, 5, 6]

Whenever you need to take a set of lines or curves and turn them into tubes, use this plugin. It is even possible to use it on branching lines as in, for example, a truss. You can specify parameters such as diameter, precision, and even material.

A similar plugin that also allows for an inner diameter to be set is **PipeAlongPath** by TIG [5,6]. It won't work well with branching lines, though.

Make Fur

By tak2hata—Free [5, 6]

This is a fun and very useful plugin. It allows you to place a large number of randomly scattered objects onto faces. These objects could be some of the default shapes (rectangles, grass blades, leaves) or any component in your model.

You have full control over the plugin's options, such as how many to place per unit area, maximum objects per face, and their scattering (randomness of scale, rotation, etc.). You can also apply a "force" parameter that distorts each object similar to bending blades of grass or hair in wind.

This plugin begs to be experimented with. Common usage includes placing grass on ground (as individual blades or PNG materials with transparency) and trees on hills, or creating any other "furry" object, such as a shaggy carpet.

Another plugin that allows you to "spray" components onto any surface is **Compo Spray** by D. Bur. It is free, and you can download it from [3].

MakeOrthoViews

By J. Folz—$5 [4]

This plugin allows you to take a 3D object (a group or a component) and create standard engineering views (front, right, and top) as additional 2D objects in your model. This is very useful if you want to export them into 2D CAD software for annotation.

Mirror

By TIG—Free [5, 6]

This plugin does exactly what its name implies: It adds the missing mirror tool to SketchUp. You can use it on 2D shapes or 3D objects.

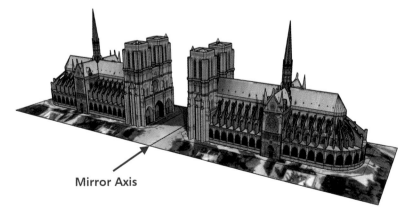

Mirror Axis

PathCopy

By R. Wilson—Free [4]

Whenever you need to copy objects along a path (rather than orthogonally), you need to use this plugin. It places a component regularly along a curve either at its vertices (endpoints of line segments) or at specified distances. The image here shows streetlights copied along a spline at 10' increments.

Pen Tool +

By R. O'Brien—Free [5, 6]

This plugin adds a toolbar from which you can use various pen-based functions. They help you modify meshes manually and add soft lines, hidden lines, guidelines, and others.

Randor

By TBD—$3 [4]

This plugin randomly rotates and scales a selection of components. This is especially helpful for making landscape objects like trees look more natural, because when they are initially placed in SketchUp they all appear with the same size and orientation. The "Randomizing Everything" example in Chapter 6 creates a similar result.

Round Corner

By Fredo6—Free [5, 6]

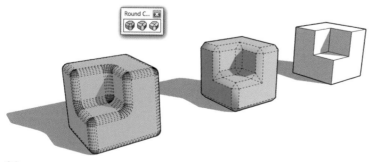

The lack of classic CAD-type chamfer and fillet tools in SketchUp has spurred several plugin creators into action. While there are also 2D versions available that can operate on a face or polylines, this 3D-version cleanly rounds (fillets) or breaks (chamfers) all connected edges in your model. Use it to give materials a more realistic appearance for rendering (no edge in real life is as sharp as a computer-produced edge) or to connect pipes realistically.

Selection Toys

By thomthom—Free [5, 6]

This plugin adds a toolbar containing a multitude of tools to aid in selecting objects in SketchUp. It works very much as a selection filter and allows you to include or exclude certain categories (faces, edges, components, etc.) from the current selection. In addition, you can select connected edges or faces and similar components. This can even be done by layer.

As a bonus, this plugin lets you customize the number of buttons that are shown—a very nice user interface feature.

Shape Bender

By C. Fullmer—Free [5, 6]

When you do woodworking, you sometimes need to bend a shape based on a curve. This plugin lets you do this virtually. Add a group or a component, a reference line, and a curve (or a set of curves). As you can see in the image shown here, everything is bent along the curve shape.

Soap Skin & Bubble

By J. Leibinger—Free [www.tensile-structures.de/sb_software.html]

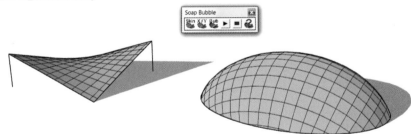

Based on the theory of tensile structures (fabric or canvas roofs), this plugin takes a set of edges and stretches a surface between them. With this plugin, it is quite easy to create hyperbolic paraboloids (hypar shells) as well as surfaces that are held up by internal pressure (as in air-supported structures).

To use this plugin, select a set of edges and fit a surface with the first tool on the toolbar. If you want to add pressure to the surface (to blow it outward or suck it inward), use the third tool on the toolbar.

SubdivideAndSmooth and Artisan (SDS2)

By Whaat—$15 and $39 [4, www.artisan4sketchup.com]

This has been the organic-modeling plugin of choice in SketchUp for a while now. What started as the SubdivideAndSmooth plugin has lately become the expanded Artisan plugin. While SketchUp may not be known specifically for polygon-based organic modeling, it is very much capable of creating non-rectilinear shapes, as this plugin shows.

With this plugin, you can subdivide any object and then crease, push/pull, sculpt, paint, and otherwise deform and change polygons to your artistic liking.

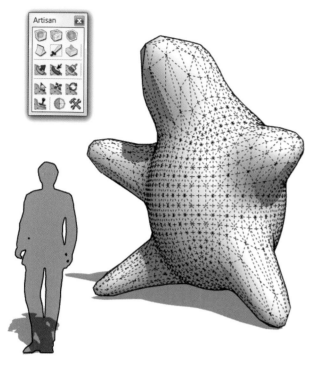

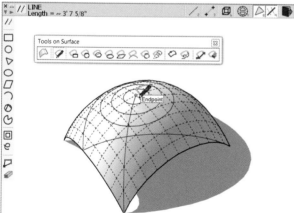

Beyond using it for shaping humanoid and animal-shaped blobs, you can benefit from its capabilities to create rocks, beanbags, jewelry, and much more.

Another plugin that lets you freely deform any shape (based on an encircling grid of points) is the **SketchyFFD** plugin by C. Phillips. It is free, and you can get it from [5,6].

Sculpt Tools by BTM also lets you modify meshes by sculpting. It is free, and you can download it from [5, 6].

Vertex Tools by thomthom is another commercial plugin ($20) that helps edit vertices on objects such as surfaces. You can select vertices smoothly and then modify them. You can get it from www.thomthom.net/software/vertex_tools.

Tools on Surface
By Fredo6—Free [5, 6]

When you work with surfaces, at some point the need will arise to "draw" on them. One example is creating a line that needs to connect two points on the surface but can't leave the surface (a so-called geodesic line).

This plugin allows many other geometric shapes to be drawn on a surface. The accompanying image illustrates this by showing two lines extending from edges of the surface on one side to the other side's midpoint. You can also see how easy it is to create circles on a surface.

You can use this tool to section a surface for application of different materials. You can also use it to help you construct shapes that are defined by the surface's geometry.

Weld
By R. Wilson—Free [4]

To be able to use some tools in SketchUp, lines need to be connected into a polyline. This allows all of the lines to be selected and modified as one object. This plugin adds the ability to "weld" many individual (attached) lines into a polyline.

Table 4.1 lists some additional plugins for modeling with SketchUp.

Table 4.1: Some Other Useful Modeling Plugins

Name	Description
Fredo Tools (Fredo6)	Various smaller tools for editing [5, 6]
ExtrudeAlongPath (TIG)	Extrudes a shape along a path [3]
Fillet (Cadalog)/Chamfer Along Path (A. Joyce)	Two plugins that add these missing features [3]
MakeFaces (T. Burch)	Creates faces where possible [4]

Name	Description
Nudge (T. Burch)	Uses arrow keys to move objects [4]
Random_pushpull (TIG)	Randomly pushes/pulls faces [5, 6]
Bezier Patch (ngon)	Creates a Bezier patch [5, 6]
Edge Tools (thomthom)	Simplifies curves, cleans up collinear edges, and divides faces [5, 6]
Guide Tools (thomthom)	Places construction points easily [5, 6]

Example 4.1: Creating Trusses

In this example, we'll create two trusses made of different materials: steel and wood. This allows us to use the **Lines2Tubes** and **Extrusion Tools** plugins as well as one of SketchUp's new Solid Tools. Make sure you have both plugins installed before you proceed.

1. Set up a grid with your desired spacing using the **Tools → Grid** menu. You might need to install the Grid plugin first—see Chapter 3 for instructions. In this example, I use a 4' grid to lay out the truss.

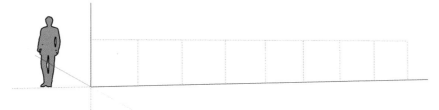

2. Draw lines at the centerlines of all members. SketchUp fills in the areas, but you can delete those faces (see next step).

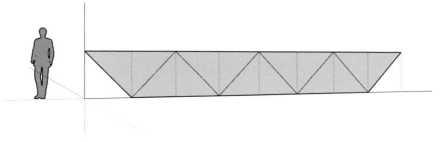

3. Highlight all of the created faces and delete them.

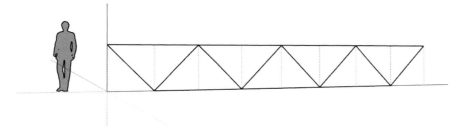

4. Select all of the lines (the axis centerlines in the truss). In the Tools menu, select the item "Convert arcs, circles . . .," after which the dialog shown here appears.

5. Set the parameters as shown in the image. As you can see, you can enter the diameter, the circle precision, and the material. We'll make sure "Each tube is a group" is selected so that we can use those groups later. Then just click on OK. The plugin gives you a confirmation, and your truss should resemble the image shown here.

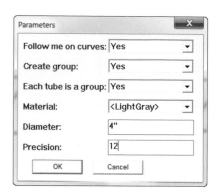

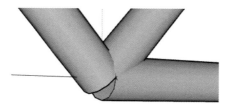

6. When you look more closely at the truss, you will see that it needs some cleaning. The ends don't extend (or at least don't clean up nicely) as would likely be the case in reality. Also, internal ends overlap too much, creating a lot of internal geometry.

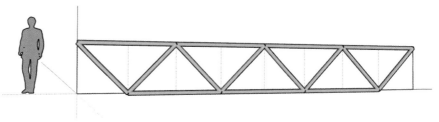

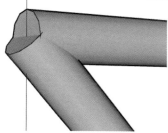

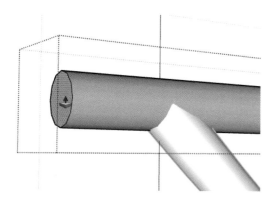

7. To clean up the ends, double-click the horizontal tubes to get into group editing mode. Then extend each of them by a reasonable number (8 inches, in this case).

8. At this point, your truss looks better, but the geometry is still overlapping quite a bit. You can see this when you turn on X-ray mode.

   Now you can do one of two things: The classic SketchUp method to clean up overlaps is to actually explode the entire truss and use the Intersect with Model right-click menu item on the entire truss. This creates lines where faces intersect, which would let you clean up overlapping faces easily.

For this example, we instead use one of the new Solid Tools that were released with SketchUp 8. While most of them are reserved for the Pro version of SketchUp, fortunately for us, the Outer Shell tool is included in the free version, too. You can find it under the Tools menu (or on the Solid Tools toolbar).

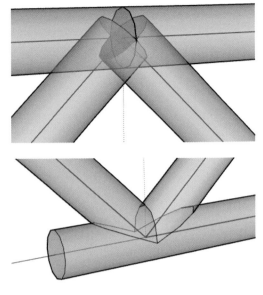

9. We can use the Outer Shell tool only on "solid" (or "watertight" objects without holes). Because we created separate groups for each tube, we actually have a model that consists only of those. Select the entire truss (select the groups that make up the members) and ask SketchUp to perform the Outer Shell tool on it. Intersections will then resemble the image shown here.

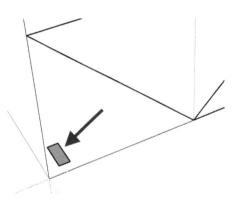

As you can see, intersections are now cleanly mitered and all internal geometry has been removed. The steel tube truss is finished!

Now let's try this with a wood truss. Obviously, we won't be able to use circular cross sections this time; therefore, we'll use the Extrude Edges by Face tool from the Extrusion Tools plugin.

1. Start with the centerlines as before (you can move away the truss we just created; the centerlines will still be in their original position). Add a rectangle near them on the ground plane (the x-y plane). This rectangle must have the cross-sectional dimensions of the wood members we want to use.

2. Select the rectangle and all of the lines that make up the truss. Then click on Extrude Edges by Face on the Extrusion Tools toolbar.

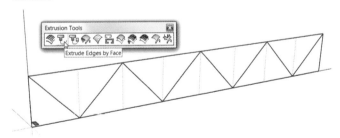

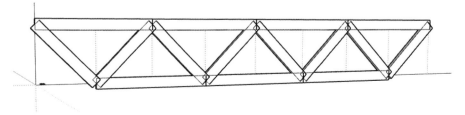

3. In the dialog that follows, say Yes to "Process extrusion in pieces." As you can see in the accompanying image, the truss has again been created as a collection of individual members (as groups).

4. Now use the same process as before to clean up the truss and make it a solid piece. The final result should resemble the image shown here.

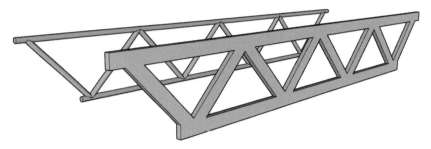

Example 4.2: Making Shells in Different Ways

In this example, we'll use the **Extrusion Tools** and the **Soap Skin & Bubble** plugins to create various shapes of shells and fabric structures. Make sure you have both plugins installed before you proceed.

First, we'll create a *circular shell*:

1. Draw two vertical arches that are at a right angle to each other: one on the x-z plane (between the red and blue axes) and one on the y-z plane (between the green and blue axes).

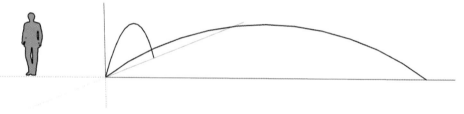

2. Now turn each of them into a group by highlighting each separately and selecting **Make Group** from the **Edit** menu. The result should resemble the image shown here.

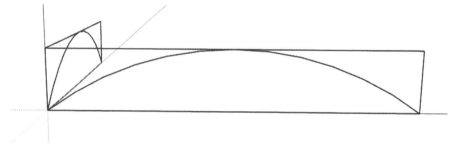

3. With both groups selected, click the Extrude Edges by Edges tool from the Extrusion Tools toolbar. In the set of dialogs that pops up, ask the plugin to orient the faces, not reverse faces if white is up already, not intersect with self, erase coplanar edges, not triangulate any faces, and keep the original groups. The result should resemble the image shown here.

You could have done this in many other ways, too. If you had created four guiding arches, then the Extrude Edges by Rails method would have worked. In addition, the Skinning tool from the **Curviloft** plugin would lead to the same result.

Next, let's create a ***hypar shell*** (a **hy**perbolic **par**aboloid):

1. Draw a square (30′ × 30′, for this example) and delete the internal face so that you are left with only edges, as shown here.

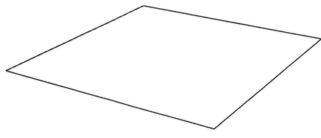

2. Move a corner point upward (by 10′ along the blue axis) using the Move tool.

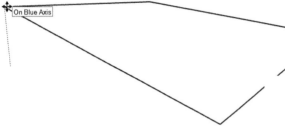

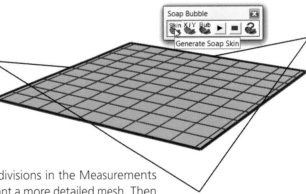

3. Repeat this with the corner point on the opposite side. You should now have two opposing edges "on the ground" and the other two "in the air."

4. Now select all four edges and click on Generate Soap Skin from the Soap Skin & Bubble toolbar. This adds a temporary set of faces into your edges.

5. You now have the option to modify the number of divisions in the Measurements Box. The default is 10; enter a different value if you want a more detailed mesh. Then hit the Return key and watch how the plugin fits the surface between your edges.

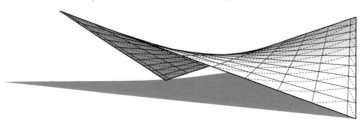

This shell will be executed as a *lofted shape*. For this, we need the **Curviloft** or the **Extrusion Tools** plugin.

1. Create three parallel, vertical arches as shown here.

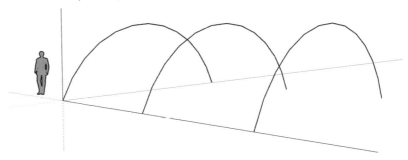

2. Now scale them so that they all have different heights and widths.

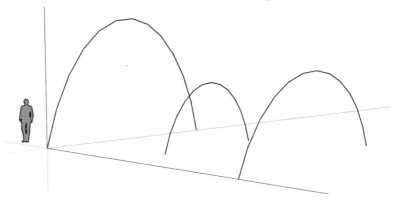

3. Highlight all three curves and select the first tool (the Loft tool) from the Curviloft plugin's toolbar. This creates a temporary shell on which you can still make modifications (move vertices, for example).

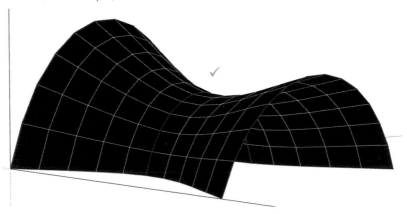

4. Once you are satisfied with the shell's shape, left-click the mouse to accept. You should now have a shell that is smoothly defined by the three guiding arches.

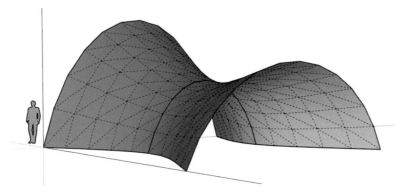

It is actually quite interesting to see that if you use the Extrude Edges by Loft tool from the Extrusion Tools plugin for this, you will get a different lofted shape, which resembles the image shown here.

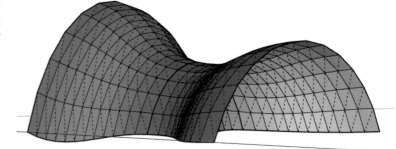

Plugins for Architectural Modeling

1001bit Tools
By Goh Chun Hee—$29 [www.1001bit.com]

This is a large collection of architecture-related tools. It currently comes in a standard and a Pro version and even has a shadow-analysis tool as an add-on. Tools vary from general modeling helpers (fillet, chamfer, arrays, etc.) to tools for creating walls, openings, stairs, floors, roofs (including rafters and purlins), and doors and windows.

BuildingStructureTools
By tak2hata—Free [5, 6]

This plugin helps with creating beams, columns, slabs, and the like. You can customize the database of shapes that it ships with and add your own profiles.

HouseBuilder
By S. Hurlbut, D. Bur—Free [3]

If you want to model each stud, rafter, and joist in a light-framed house, then give this plugin a try. It comes in both metric and imperial versions and adds a toolbar that gives you easy access to all of its tools.

You can use the Wall tool to create a stud-framed wall based on two selected points. To add a window or a door, select the appropriate tool and pick a location along a wall. The

plugin places the opening and frames the wall around it. To add a roof (with evenly spaced rafters), select two points on the gable end and one point on the opposite side. The plugin then creates rafters—complete with birdsmouth and eave cuts.

Plan Tools
By thomthom—Free [5, 6]

This set of small tools was written to make work with DWG imported data easier. You can, for example, use an imported file with building outlines to create the massing for an entire city in just a few clicks. Another tool moves all selected geometry to the ground (z = 0) or onto Layer0.

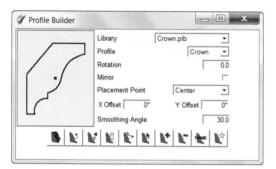

Profile Builder
By Whaat—$29 [4]

This plugin picks up where the Follow-me tool leaves off. It allows you to create an extruded profile without having to create a guiding edge first. It can also apply extruded cross sections to edges in your model.

The current selection of cross-section shapes includes millwork, road curbs, concrete shapes, and an international collection of structural steel shapes. You can preview the shape and apply various parameters when placing them.

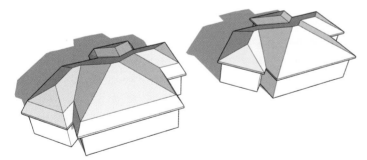

Roof
By TIG—Free [5, 6]

If you don't want to create a complex roof yourself using extrusions, intersections, and quite a bit of cleanup, then download this tool. It automatically creates clean hip, mansard, gable, and pyramid roofs for you. As you can see in the image shown here, even angled floor plans work with this plugin.

SketchUp Ivy
By Pierreden—Free [5, 6]

This is a great plugin if you want to quickly add foliage to an architectural scene. By default, it creates an ivy stem that grows along a face or a set of faces. You can control branching, scattering, and direction parameters, but, due to the random nature of the algorithm, you will get different results each time you run this plugin.

I can imagine this plugin working equally well for gnarly bushes and for vines—all you need to do is provide appropriate foliage. You accomplish this by adding to the model leaf components that have "ivyleaf" in their name.

For good results, we need to edit the foliage materials that come with the plugin. Export the leaf images to your image-editing software, remove the background (make it transparent), and then reimport into SketchUp's materials as a transparent PNG.

Tools for Architects

*By rubits—Free/various prices [*www.rubits.com*]*

These tools help with massing, textures (materials), and selections. The Massing Box tool allows you to draw rectangular shapes with a defined (fixed) area, and the Area Scale tool lets you scale any shaped face to a defined area. Both of these tools are excellent help in the programming layout phase of planning.

Windowizer

By R. Wilson—Free/$10 [4]

If you need "storefront" mullion-type windows in your model or want to panelize a surface, then this plugin can create them for you based on a collection of selected faces. You can specify mullion width and a variety of other parameters. The plugin then places mullions at each face edge and replaces the face's material with a transparent glass material.

To use this plugin, select one or more faces and then right-click on one of them. The Windowize command is in the context menu.

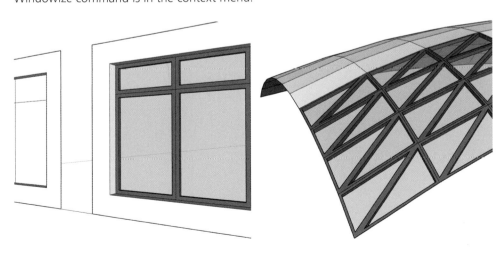

Example 4.3: Mullion Windows

In this example, we use the **Windowizer** plugin to create storefront windows for an otherwise quite plain concrete building. Make sure the plugin is installed before you continue.

1. Start with a box. In this example, we use a 30′ × 30′ square, which we pulled up by 15′, and to which we then applied a concrete texture.

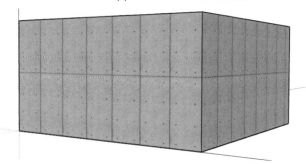

2. Create locations for the windows. For this example, we offset the front facade by 2′ and then pushed it inward by 1′.

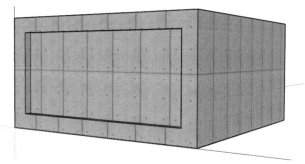

3. On the facade rectangle, draw lines wherever you want to have mullions. This breaks the large face into many small ones.

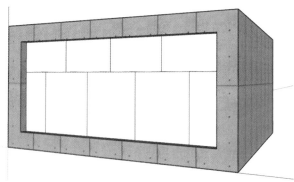

4. Now highlight all of the faces that will be windowized and right-click on one of them. In the context menu, you will find the Windowizer tools. Click on the Windowizer tool menu item to get to the settings dialog.

5. Adjust these settings to your liking. As you can see, you can even pre-select the materials for the frame and the glass. When you click on OK, your windows will be generated for you.

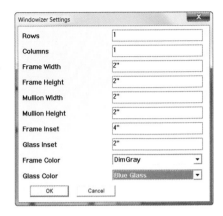

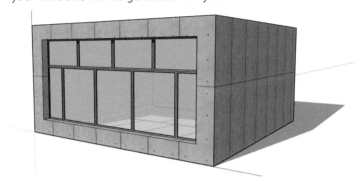

Plugins for Digital Fabrication

CADspan Tools

By CADspan—Free and Pro [www.cadspan.com/software]

In addition to providing 3D printing services, this company has published a free plugin for SketchUp that allows you to detect problems in a 3D model and provides a resurfacing tool that produces 3D-printable "watertight" meshes. You can then export your model as an STL file.

Center of Gravity

By TIG—Free [5, 6]

Given a solid shape, this plugin finds the 3D center of gravity of the shape and marks it. If you do this for more than one shape, it can also calculate the combined center of gravity.

This plugin can help with sculpture as well as mobile making.

Centroid and Area Properties

By A. Schreyer—Free [7]

This plugin finds the centroid of any face that is drawn on the ground plane. The centroid is marked, and several relevant properties such as Area, Moment of Inertia (in two main directions), and Radius of Gyration are reported.

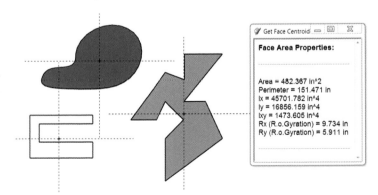

103

Cutlist and Layout

*By S. Racz—Free [*http://steveracz.com/joomla/content/view/45/1*]*

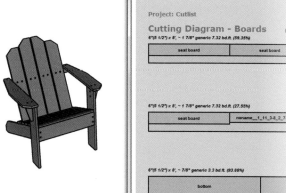

This excellent plugin is geared mainly toward woodworkers but may be useful to others as well. It takes a model that consists of parts (components such as individual wood pieces that make up a table) and produces materials lists as well as cut sheets (with layout optimization and efficiency reporting). The user can separate solid wood, sheet goods, and hardware simply by adding keywords to the components.

DrawMetal

*By Draw Metal—Free [*www.drawmetal.com*]*

This website, geared toward metal fabricators, provides a set of three free SketchUp plugins (Curve Maker, Taper Maker, and Stock Maker) that allow you to create intricate bent metal shapes for railings and other projects. The Curve Maker tool makes it easy to create many beautiful spirals and curves such as the Archimedes spiral, a catenary, a helix, a golden spiral, and others.

Flatten

By T. Burch—$10 [4]

This plugin allows you to flatten a group or a component by arranging all faces on the ground plane.

Flattery Papercraft Tools

*By pumpkinpirate—Free [*www.pumpkinpirate.info/flattery*]*

This plugin lets you unfold objects similar to the Unfold plugin. It has interactive features such as adjusting the unfolded shape and adding glue tabs.

Although not a SketchUp plugin but rather an external software, **Pepakura Designer** should be mentioned here, too, because it can take a SketchUp model exported as a COLLADA DAE file and unfold it. *$38* [www.tamasoft.co.jp/pepakura-en]

Glue Tab

*By rschenk—Free [*https://github.com/rschenk/Sketchup-Glue-Tab-Plugin*]*

This plugin makes for a great companion tool to the Unfold tool. It allows you to select edges in an unfolded model and then adds glue tabs to them.

i.materialise

*By i.materialise—Free (+ printing costs) [*http://i.materialise.com/support/plugins*]*

This plugin allows you to take a 3D model and preprocess it for 3D printing through the i.materialise website.

Manifold

By TIG—Free [5, 6]

To be able to 3D-print an object in SketchUp (or use any of SketchUp's Solid Tools), it is important that the object consist of manifold geometry. This means it cannot have gaps or disconnected geometry. This plugin tries to fix grouped geometry and gives you errors if it can't do so.

TIP

You can easily check whether SketchUp sees your group or component as a "watertight" manifold by looking for the volume entry in the Entity Info window. If there is no such entry, then you may have gaps in your object.

Outer Shell and Solid Tools

By SketchUp—Included with software

Technically, this is not a plugin, since all the tools are included in SketchUp. They do, however, need to be included in this list.

Outer Shell takes several intersecting solids and deletes all interior geometry (creating a "watertight" manifold shape). It also produces clean intersections. You can use this tool (which is included in the free SketchUp version) to preprocess your geometry for 3D printing.

The other Solid Tools (Intersect, Union, Subtract, Trim, and Split) are also very useful in modeling for fabrication; however, they are included only in the Pro version of SketchUp.

Phlatscript

By Phlatboyz—Free [www.phlatboyz.com, 9]

This plugin allows you to preprocess geometry in SketchUp and export it to G-Code for CNC (Computer Numerically Controlled) milling. It was developed by the makers of the Phlatprinter for that machine, but the code might work with other CNC mills as well.

Polyreducer

By Whaat—Free [5, 6]

Although the author claims that this plugin is not perfect, it is a good tool to have if you need to reduce face count in any polyface mesh.

Slicer

By TIG—Free [5, 6]

When you have an object that encloses a volume (such as a blimp or a sculpture), then one manufacturing method is the egg-crate approach, whereby the object is sliced up into evenly spaced thin sheets in two main directions. You can then cut the 2D sheets and reassemble them into the "core" of the original shape by sliding each one into overlapping slots.

This plugin allows you to egg-crate any geometry. Its options let you specify dimensions and spacing of the slices as well as other parameters. After the slicing operation, all slices can be neatly arranged on the ground plane (for printing or exporting). If you have the Pro version of SketchUp, a DXF export option is included. In the free version, use orthogonal view printing (or a free DXF exporter plugin) to create scaled output.

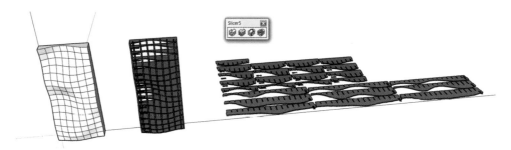

Solid Inspector
By thomthom—Free [5, 6]

This is a useful plugin that checks groups and components to determine whether or not they are "solids." If errors (e.g., holes) are found, the plugin circles them. This helps prepare geometry for 3D printing.

TF Rubies
By C. Bremer—Free [www.timberframecottage.com/su]

Geared for timber framers, this plugin allows a frame designer to either use components (such as beams, posts, joists, and braces) from a library or make their own. Bents and entire building frames can be assembled quickly using these components (including all connections and cuts), and frame members can be individually exported for multiview cut sheets, as you can see in this image.

Company Name
Project: TF_Ruby Example
Timber: Girt III-IVA - 8.0 x 8.0 x 16
Created on: 01/24/2012

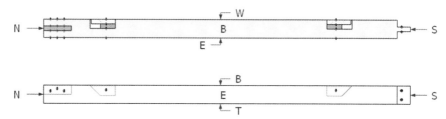

Unfold

By J. Folz—Free [10]

This plugin is an excellent helper if you need to take 3D geometry from SketchUp and "flatten" it into 2D for cutting and manufacturing. This method works very well, especially for thinner materials like paper or sheet metal.

By itself, this plugin does not add glue tabs to the flattened sheets; use the Glue Tab plugin (previously discussed) for those.

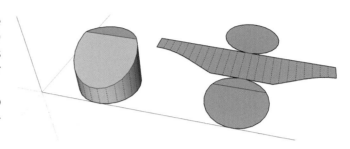

Waybe

By Waybe—Various costs [http://waybe.weebly.com]

This commercial plugin provides an all-in-one solution for unfolding objects (including flattened layout and glue tabs).

Wudworx Tools

By Wudworx—Various costs [https://sites.google.com/site/wudworx]

Also geared toward woodworkers, this site currently offers three plugins: Board Maker, Mortise & Tenon, and Dovetails.

Example 4.4: Slicing Up a Volume

This example shows how you can use the **Slicer** plugin to prepare a solid for egg-crate-type manufacturing. To make things easy, we'll use one of SketchUp's default spheres (get it with the Component window) for this. However, the plugin will work on any solid group or component (it just must be "watertight" and without holes). Make sure you have the plugin installed before you begin.

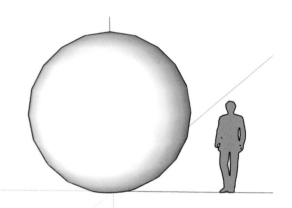

1. Highlight the sphere and select **Slicer5** from the **Plugins** menu (you can also use the toolbar for this). This opens up a dialog with options.

2. For this model, we want to slice the sphere into vertical and horizontal planes. Therefore, select "XZ" as the axis combination. Also, set all other parameters as you see them in the dialog image.

3. Because we are slicing this sphere in two directions, the plugin asks us for the second axis parameters in another dialog.

Slicer5: Parameters

Axis:	XZ
Spacing:	12"
Thickness:	1"
Inset at Start:	6"
Inset at End:	6"
Add References ?	Yes
Text Height:	2"
Flatten ?	Yes

OK   Cancel

Slicer5: 'Z' Slices

Spacing:	12"
Thickness:	1"
Inset at Start:	6"
Inset at End:	6"
Slot ?	Yes
Tolerance:	0"
Overcut:	0"

OK   Cancel

4. After clicking OK, the plugin gets to work. Be patient at this point because this could take some time. You can follow its progress at the bottom of the screen, if you like. When everything is done, you should see something that looks like the image shown here.

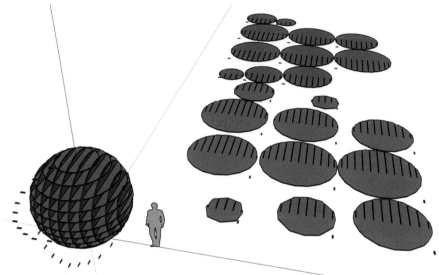

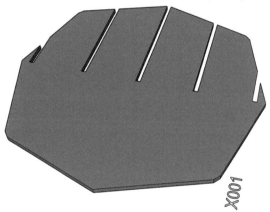

5. As you can see, the sphere has been sliced evenly into planes in both the horizontal and the vertical direction. In addition, copies of all of these planar elements have been placed onto the ground next to the sphere, ready to be printed or exported. Everything is color-coded and neatly labeled. The image here shows a close-up of one of the slices.

You can now export the slices as DXF shapes if you have the Pro version of SketchUp (or if you have a DXF exporter plugin installed). If you instead want to print the shapes, switch the camera mode to Parallel Projection and go into Top view before printing (you can do both in the **Camera** menu).

Example 4.5: Making a Terrain Model

A common task for architects and landscape designers is to create a 3D site model. This is often done by slicing up a site horizontally (at equally spaced elevations), fabricating contoured slices, and then assembling the terrain model. We again use the **Slicer** plugin to accomplish this task.

1. A benefit of using SketchUp is that we can get any terrain imported into our model through SketchUp's location tools. Unless Google's terrain model is too coarse or inaccurate, there should be no need to find alternate data sources. To locate your model, click on the "Add location . . ." tool on the main toolbar.

2. This brings up a dialog where you can browse to your site. Once you have the location in view, use the Select Region button to select an appropriate region.

3. This process imports two items into SketchUp: a flat orthographic view of your site and a textured terrain model (both are correctly georeferenced and oriented). You can switch between them by toggling visibility for the appropriate layers, as shown here.

4. Turn on the terrain model visibility. Next, add a rectangular box where you want to acquire site data. Make sure they overlap, as shown here.

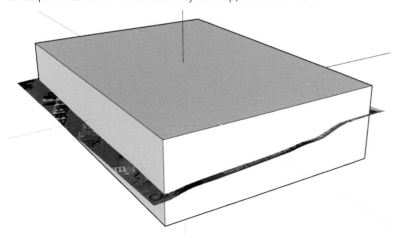

5. Select the terrain model. As you can see, it is highlighted by a red box, which means that it is locked. Right-click on it and choose the Unlock menu item. This makes the terrain editable (it will also get a blue outline).

6. Right-click on it again and choose Explode. This removes the terrain from its grouping.

7. Now *triple*-click the box you created earlier and select Intersect Faces with Model from the right-click menu.

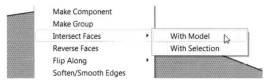

8. This intersects the two objects so that you just have to remove the top portion (simply erase the lines and faces) and you will be left with a clean solid terrain model.

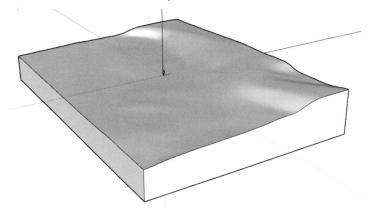

9. You could leave the terrain as is, but, ultimately, we don't need the thickness below the terrain. Therefore, push the underside up as far as it will go.

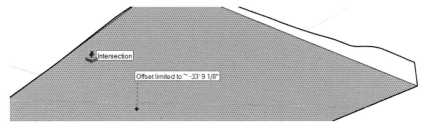

10. Group the site object first so that you can start the Slicer plugin. Go with the settings shown here (we want to get horizontal slices at 4' intervals).

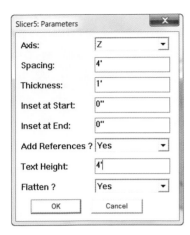

11. This slices up the site and also creates flattened slices that you can again print/plot or export to a laser- or CNC-cutter, for example.

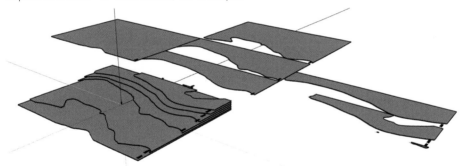

Example 4.6: Unfolding a Shape

In this example, we use the **Unfold** plugin to unwrap a solid shape for manufacturing. The shape we use here is a circle extrusion that received an angled planar cut.

1. Start by creating a shape. This does not have to be a solid; surfaces also work well.

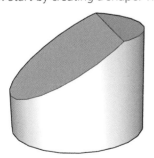

2. Because we will be unwrapping a circular shape, we need to see all hidden lines during the process. Go to the **View** menu and make sure Hidden Geometry is turned on. You should now see the dashed lines, as shown here.

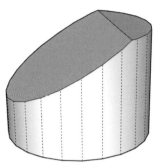

3. With nothing selected in your model, go to the **Plugins** menu and click on **[JF] Unfold**. To make the first unfolding, click first on the face you want to unfold and then on the face that has the orientation you want the first face to acquire. In the example shown here, do the clicks in the order shown to fold the angled face upward.

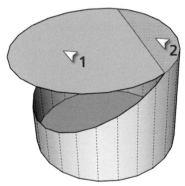

4. Now we need to fold both of those faces so that they are parallel with the center face on the rear of the object. First, click in white space to unselect the last face. Then shift-click both top faces to select them. Finally, let go of the Shift key and click the back-side face.

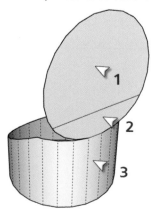

5. Again unselect everything by clicking in white space, and turn to the front of the object. Start in the center and unroll the sides by clicking on consecutive faces (stop when you match up with the back side).

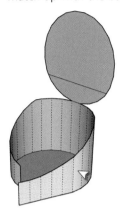

6. Your object should look like this after you unroll both sides.

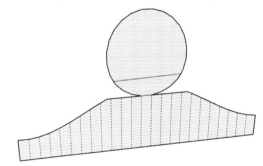

7. As a final step, flip the base backward, too. Then you can exit the tool.

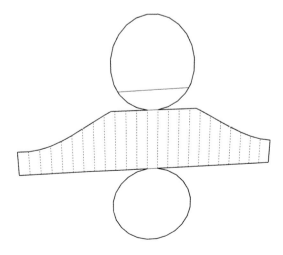

You can now rotate this shape onto the ground plane and switch your view to a parallel projection from the top. At this point, you are ready to print the unfolded shape scaled and use it as a template for cutting the shape out of, for example, sheet metal.

After you perform this process with different shaped surfaces, you will notice that some surfaces can be easily unrolled (as in our example) while some others cannot. A surface that is unrollable is generally called a *developable* surface. Examples are cylinders, cones, and any ruled surface that was derived using a line. Double-curved surfaces (like the shells shown earlier in this chapter) cannot (easily) be unrolled using this method. You may be able to unroll them in strips, though.

Plugins for Data Integration and Exchange

Cloud
By D. Bur—Free [3]

You can import x, y, z coordinates into SketchUp from a text file using this plugin. Coordinates get added as construction points, and the resulting point cloud can be triangulated after import (to create a surface). This can be useful, for example, if you have LIDAR laser-scanned point data that you want to use as a basis for building modeling. Be careful with LIDAR data, though, as this plugin slows down with a too large number of data points.

As a bonus, existing vertices (edge points) of any highlighted geometry can be exported to a CSV file.

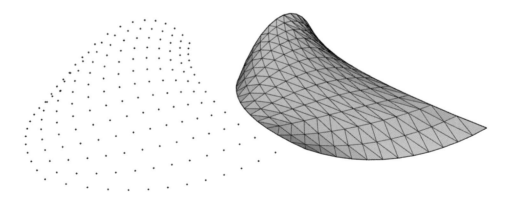

A commercial plugin, **Pointools**, claims to be able to handle large numbers of LIDAR data points easier in SketchUp. (Free evaluation, $1350 for full plugin [www.pointools.com].)

TIP

If you want to process LIDAR data files (that typically come in the native LAS format) for use with SketchUp (as x, y, z formatted CSV files), then the freely available **LAStools** software might be of help:

www.lastools.org

Import/Export Plugins
By Various—Free [3, 5, 6]

Currently, there are many limited-feature import/export plugins available for SketchUp. If you need one for a particular file type, search the plugin repositories for the file extension (such as DXF). The most popular file formats for exchange are DXF (to exchange linework and other shapes with lots of other software including Autodesk AutoCAD), OBJ (to exchange with 3D modeling software), and STL (to send files to 3D printers).

Here is a small sampling: Dxf_In, Dxf2Skp, PLY importer, FreeDXF Importer, DXF_export, Import OBJ with Materials, OBJexporter, OBJ Import, and su2stl.rb, jf_stl_importer.

Please note: If you own the Pro version of SketchUp, then DXF and DWG import/export is included.

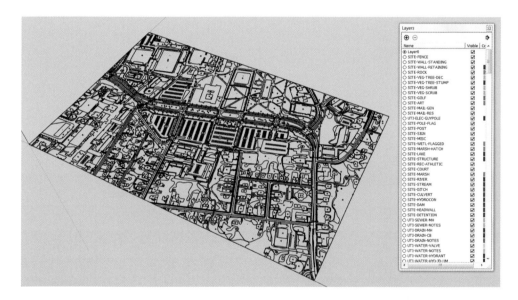

KML Tools
By Aerilius—Free [5, 6]

This tool facilitates the import and export of KML data, such as placemarks and image overlays, that cannot be exchanged with Google Earth using SketchUp's native COLLADA-based (DAE and KMZ) file formats.

When you create objects for use in Google Earth, look at this plugin to complement SketchUp's native import/export capabilities.

Layer Manager
By D. Bur—Free [5, 6, 3]

When you work with imported data, there will likely be a need to clean up layers and objects on layers at some point. This tool helps with moving objects between layers and maintaining your collection of layers.

Triangulate Points
By TIG—Free [5, 6]

This is a triangulation-only version of the Cloud plugin. You can use it to triangulate point clouds (made with construction points) that you already have in SketchUp.

Plugins for Animation and Presentation

CubicPanoOut
By J. Wehby—Free [4]

This tool takes your current viewpoint and exports a cubic panorama as a set of six JPGs. This is an easy way to create interactive 3D panoramas from SketchUp models. You can even create panoramas for models that have custom styles (e.g., in the blueprint style).

Depending on the 3D panorama format (e.g., QuickTime), it is possible to embed such a panorama on a webpage and let visitors explore a space virtually.

FlightPath and FlightPath2
By R. Wilson—Free/$7 [4]

While SketchUp's tabbed animation feature is very useful for walk-throughs and flyarounds, its results are not very smooth and the path is controllable only through a tab's viewpoint. Use this plugin to define a path and create walk-throughs based on this path.

Keyframe Animation
By Regular Polygon—$24 [http://regularpolygon.org/keyframe-animation]

A commercial object animation plugin. You use it by assigning animation tabs in SketchUp where you want keyframes to be. This plugin saves object positions with animation tabs and animates objects (linearly and by rotation) when you switch tabs or play the scene animation. Animations can also be exported to a movie.

Mover
By Unknown—Free [5, 6]

This small plugin lets you add animation to SketchUp. Unfortunately, it is not well documented (and is hard to find), but it works well.

PageUtilities
By R. Wilson—$5/$7 [4]

This set of utilities makes working with SketchUp's tabbed animation feature easier by providing smooth transitions and constant speed.

Proper Animation
By morisdov—Free [5, 6, https://sites.google.com/site/morisdov*]*

This plugin allows you to create animation in SketchUp by assigning positions to objects and animating them through SketchUp's tabbed animation.

SketchyPhysics
By C. Phillips—Free [5, 6]

SketchyPhysics has two amazing purposes: (1) simulating physically accurate motion behavior (including collisions) in SketchUp and (2) playing bowling or dominos right in SketchUp. This plugin incorporates a physically accurate simulation engine into SketchUp that comes from game software. When it is running, it is capable of simulating gravity, bounces, drops, and even buoyancy—you can interact with the model, too, by simply dragging on objects. A good example is pulling a bowling ball toward some pins that fall over on impact (see the images that follow for examples).

One use for this plugin is to create realistically piled-up objects for rendering. To do this, group each object separately and place them all above the ground plane that SketchyPhysics provides. Then run the simulation and watch your objects fall and tumble over each other. Save a copy of the SketchUp file to preserve the geometry before you exit SketchyPhysics.

Another plugin, **SketchyReplay**, allows you to record this motion in SketchUp, which can later be used for rendering animations.

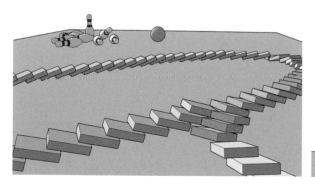

Smoothstep Animation
By Whaat—Free [5, 6]

This plugin smoothes out transitions between view tabs in SketchUp's scene animation feature.

Stereo
By TBD—$8 [4]

This plugin saves two anaglyph images for the current view that you can use to create a stereoscopic view of the current scene.

SU Animate
By Cadalog—$88 [www.suanimate.com]

This commercial plugin allows you to animate objects and camera movements in SketchUp using keyframes (animation tabs) or paths. It also has a companion product, **SU Walk**, which provides photorealistic animation for SketchUp models.

VisMap
By M. Rinehart—Free [5, 6]

If you use SketchUp's tabbed animation to hide elements during transitions (for example, to create assembly animations), then this plugin can simplify your workflow. It lets you select each layer's visibility for the tabs in your model using a single dialog.

Zorro2
By Whaat—Free [5, 6]

When you need to cut a model at an arbitrary plane (e.g., to create section renderings for a house model), this plugin facilitates this process.

Example 4.7: Making a 3D Interactive Panorama

In this example, we will create a 3D interactive panorama from a SketchUp model using the **CubicPanoOut** plugin and GoCubic, a free panorama stitching software. Use any model you like, and set up the visuals using the Styles window before you start. Also, make sure you have the CubicPanoOut plugin installed.

1. Navigate in your model to the viewer's location that you want to export as the 3D panorama. Check that the location is giving you the desired viewpoint using SketchUp's Look Around tool (the button with the eye).

2. To get consistent lighting for all views, make sure "Use sun for shading" is checked in the Shadows window. You can also turn on shadows, if you like.

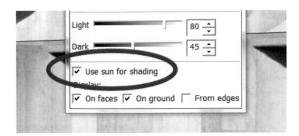

This plugin can create pages for you in SketchUp that have the correct orientation. Use them with your rendering software (they usually export as camera views) to render six (square) images that you can stitch using the instructions that follow.

3. Now that you are set up to run CubicPanoOut, go to the Camera menu and click on CubicPano Out. In the dialog that opens, set a desired resolution and adjust the options as shown here.

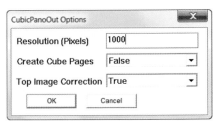

4. After the tool has finished saving all six views, you can find consecutively numbered square images, like the one shown here, in the selected location.

5. At this point, you need cubic panorama stitching software. For Windows, you can use GoCubic, which can be downloaded for free from www.panoguide.com/howto/display/gocubic.jsp. For Macs, you will need to search for similar software.

6. In GoCubic, use the "Make Pano Movie . . ." menu item to make the MOV QuickTime QTVR panorama. Just select the first exported image and the tool will use all images for stitching the panorama.

7. GoCubic then creates and displays the 3D QTVR panorama, which lets you use your mouse to look around and zoom.

If you want, you can easily embed the panorama on a webpage (search the Internet for instructions) or e-mail it to a client.

Plugins for Analysis

ColorBySlope
By Chris Fullmer—Free [4]

This plugin colors faces according to their slope. You can specify maximum color values, and it interpolates based on what you've specified. This plugin is quite useful for analyzing surface gradients. Just be aware that coloring is permanent (existing materials will be replaced).

ColorByZ
By Chris Fullmer—Free [4]

You can use this plugin to analyze heights on a surface. Based on a desired color gradient, it colors faces depending on their z value (height).

ecoScorecard

By SMARTBIM—Free (Beta) [http://ecoscorecard.com]

This online service offers a SketchUp plugin that integrates their data, consisting of green certification information for architectural and interior products, with 3D components that can be downloaded into SketchUp from 3D Warehouse. Once a SketchUp model has been populated with these components, it is possible to run an environmental audit on the entire model and report certification criteria.

OpenStudio

By NREL/USDOE—Free [http://openstudio.nrel.gov]

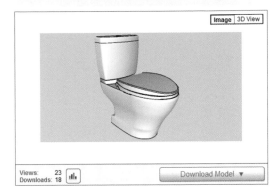

This plugin is part of the U.S. National Renewable Energy Laboratory's free suite of building energy analysis tools that include building energy modeling using their EnergyPlus software and lighting analysis using their Radiance software.

The SketchUp plugin provides tools to add data to a coarse building model and to preprocess it for analysis. This includes designating surfaces (individual faces in SketchUp) as walls, openings, glazing, interior partitions, and so forth. Material properties can be added and adjusted for the analysis. In addition, the tool features a gbXML (green building XML) importer, which facilitates data exchange with other BIM software.

This plugin has extensive capabilities for the architect or engineer interested in evaluating a building's energy performance. It is even possible to use Ruby scripting (discussed in Chapter 6) to modify this tool's features and adapt it for new uses.

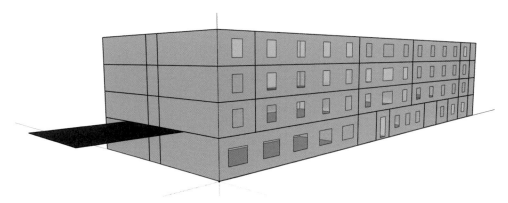

gModeller

By GreenspaceLive—Subscription [http://greenspacelive.com]

This plugin adds gbXML modeling capabilities to SketchUp (for architectural energy analysis using an analysis program of your choice). You can designate surfaces as walls, openings, and so forth, as well as add labeled spaces. It is even possible to use an autosurfacing function that will do most of this work for you. gbXML files can then be exported as well as imported.

Graphing Plugins

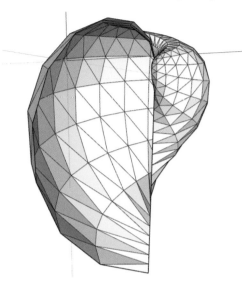

If you need to visualize mathematical functions in SketchUp either to serve as a basis for geometry creation or simply to get a 3D representation, these plugins will help you.

GraphIt—*by R. Wilson—Free [4]*—Draws 2D Cartesian equations.

k-tools—*by K. Krusch—Free [3]*—Draws 2D and 3D Cartesian, polar, and cylindrical functions. Creates faceted shells for 3D functions.

Menger Sponge—*by Regular Polygon—Free [5, 6]*—Creates a fractal geometry "spongy" box.

Sierpinski Tetrahedron—*by Regular Polygon—Free [5, 6]*—Similar to the Menger Sponge, but with triangles.

IES Virtual Environment

By IES—Free/Commercial [www.iesve.com]

IES provides a free building energy estimation and carbon analysis plugin (VE-Ware) for SketchUp as well as a commercial plugin (VE-Toolkit) that allows for quick, iterative, early-stage design analysis of a SketchUp energy model.

The commercial plugin analyzes climate, natural resources, building metrics, construction materials, energy, carbon, daylight, solar shading, water, low- and zero-carbon technologies, and ASHRAE/CIBSE heating and cooling loads. A further module can also provide Leadership in Energy and Environmental Design (LEED) credit assessment.

Example 4.8: Creating an Energy Analysis Building Model with OpenStudio

In this example, we create a small "building" (only a single room, actually) using the **OpenStudio** plugin, which could serve as a model for use in EnergyPlus building energy analysis. This is just a very basic example model. For more details and to learn about running and interpreting the analysis, consult the OpenStudio documentation. To enable the OpenStudio functionality, make sure you have the plugin installed before you begin.

When you create building models for energy analysis, keep in mind that you usually don't need to model actual material thicknesses. Walls, for example, are just single quadrilateral faces in SketchUp. Too much geometry is often counterproductive (and increases analysis time); therefore, you can usually ignore small wall recesses or protrusions.

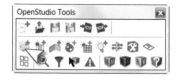

1. Before we add any walls, we need to designate a space (a zone) that will contain the area. Use the New Space tool on the OpenStudio Tools toolbar for this.

2. Click at the origin and place the new Space group. It appears as a blue outlined box.

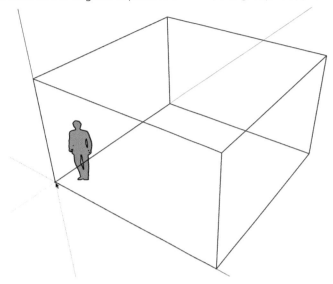

3. Now double-click it to go inside the space component (similar to SketchUp's usual group editing mode).

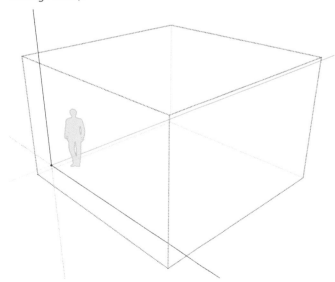

4. Using SketchUp's standard modeling tools, draw a box inside this component. This will be our building. As you can see, the outlines of the Space group adjust to the size of the geometry that you add.

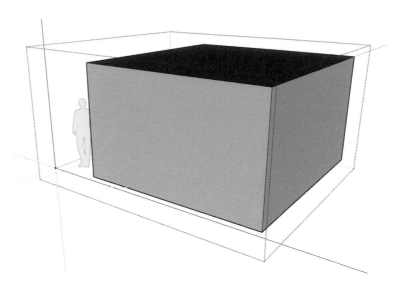

5. Without leaving the space component, draw some windows and a door onto the front surface of the box. You should notice at this point that when the door is attached to the sill level it has a different color than the windows, which are in the middle of the wall.

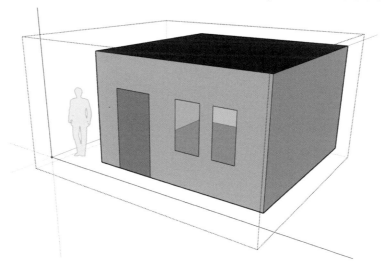

6. Let's check that we did everything right. Click on the Inspector tool on the OpenStudio toolbar. This brings up a dialog, which shows object properties when you highlight a face. Select the single face of the door and check that the dialog identifies it as a door.

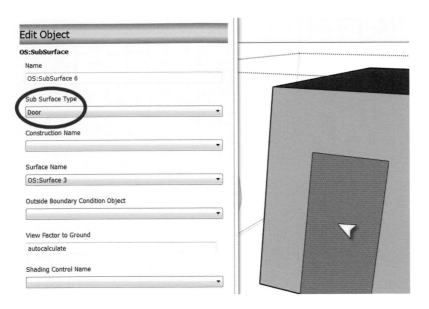

7. Also highlight the windows and the walls to see how OpenStudio classifies them.

8. Let's now add a shading device to the windows. To do this, we first need to create another group. Click on the New Shading Surface Group icon on the toolbar and place it above the windows. It shows up in the same way the Space group that we placed earlier did.

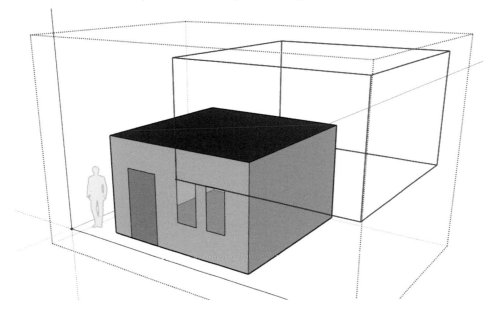

9. Now double-click the Shading Surface group to be able to add shading geometry inside. If you can't double-click it, open the Outliner window and double-click it there.

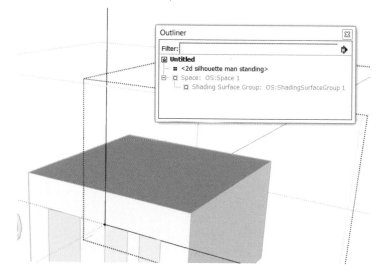

10. Finally, add any shading geometry you like. As before, keep it simple. This example simply adds a horizontal rectangle. Verify in the Inspector dialog that it is added as a "ShadingSurface."

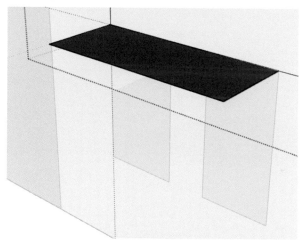

11. When you exit editing mode for both the Shading Surface and the Space groups, this is what you should see as our (very basic) finished model.

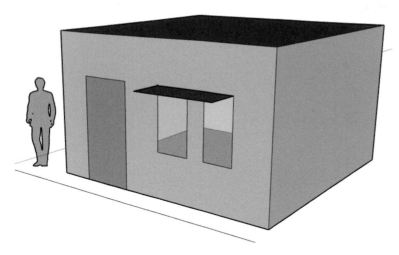

This exercise was a short introduction to setting up a building model in SketchUp for energy analysis using OpenStudio. As you can imagine, refining this model and analyzing it is more involved and therefore beyond the scope of this book. It is well documented on NREL's website, though, and I encourage you to explore this further.

Your Turn!

The following tasks allow you to practice the topics in this chapter:

1. Make a Physical 3D Model of a Building on a Site

Create a 3D model of your house or any other building you like. Use the Slicer plugin and SketchUp's location tools to create an accurate model of the site. Then make a textured 3D model of your house and use the Unfold tool to flatten, print, and fabricate it. (See **Figure 4.3** for an example and **Figure 1.2** for the finished house model.)

2. Make a Glazed Atrium Roof Shell

Model the rough outlines of an inner courtyard (e.g., of a museum). Use any of the shell tools to create a glazed shell over it.

3. From Laser Scanner to 3D Printing

If you have access to a laser scanner (or at least, to its data), load a point cloud (e.g., of a small sculpture) into SketchUp and turn it into a "watertight" solid. For a low-cost laser-scanner, give DAVID 3D a try (www.david-laserscanner.com). Then, preprocess the object for 3D printing and send it to a printer. If you don't have access to a 3D printer, give these online services a try: CADspan, i.materialise, Shapeways, or Ponoko. Some of these accept SketchUp's COLLADA files without modification. If you need to use the STL file type, look for an exporter plugin.

House

Garage   Breezeway   Chimney

Figure 4.3: Unwrapped house model, ready for fabrication

4. 3D Panoramas

Use one of your already modeled projects and create a 3D QTVR from it. Assume you are using this tool for a client presentation. As an added challenge, create the 3D panorama from a set of photorealistic renderings.

5. Walk-throughs for Clients

Get an architectural model and use any of the plugins discussed here to create a walk-through animation. Export this as a video.

6. Assembly Animation Video

Create an assembly animation in which components move. A good example is a steel connection where bolts move into place. Export this as a video.

7. Estimate your Home's Energy Usage

Using the instructions given here and tutorials from NREL's OpenStudio website, set up a very basic house for energy modeling and run an analysis. Keep it simple to begin with, so that you can thoroughly understand all the parameters.

Chapter 5
Rendering in SketchUp

This chapter introduces methods to create photorealistic renderings of SketchUp models and use them for highly polished presentation images. It covers setting up a rendering workflow, gives an overview of rendering methods, and presents techniques for setting up the rendering environment, lights, and materials.

Key Topics:

- Non-photorealistic rendering methods in SketchUp
- Principles of photorealistic rendering
- Overview of rendering software
- Using free rendering software for a sample model
- Using commercial rendering software
- Rendering components: environments, lights, materials, objects
- Tips for successful renderings
- Making renderings presentable

Let's Get Visual!

An important—arguably the most important—aspect of a designer's workflow has always been the presentation of the work. In school, this consisted of desk crits and pinups, and in work life, it encompasses presentations to clients and banks, for example, and even town hall meetings. In any of these situations, the designer's work has to be presented clearly, accurately, enticingly, and convincingly. This is the case for anything from hand-sketches through renderings to fabrication drawings.

Let's take a look at how SketchUp can help us with this.

What Is Already in SketchUp?

A major strength of SketchUp has always been its inclusion of presentation capabilities in the modeling environment. Many other CAD software packages fail to offer this combination—they typically separate modeling from presentation (think of "Model space" and "Paper space" in Autodesk's AutoCAD, for example). With SketchUp, however, it is possible to easily create a model, add annotation and dimensions, save the view of the modeling environment as an image, and then present it. It is even possible to present an annotated model interactively.

Several of SketchUp's built-in features support this workflow, in particular:

- **Font and dimension settings**—As mentioned in Chapter 2, these can be customized to the user's liking. Dimensions also display nicely and are easily readable in 3D space.

- **Shadow settings**—SketchUp has a built-in shadow feature that allows for display of accurate date- and location-based shadows. Particularly architectural models benefit from the display of shadows—surface features are enhanced by giving them the perception of depth, and buildings look more realistic when "the sun gets turned on." (See **Figure 5.1**.)

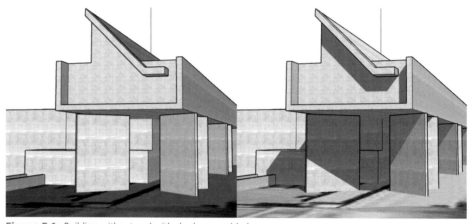

Figure 5.1: Building without and with shadows enabled

- **Fog**—The Fog window in SketchUp lets you specify a distance from the view beyond which the view's background will fade away in what looks like fog. As we will see later, this feature is actually quite useful beyond simply making your scenes look "foggy."

- **Styles**—Most customization options for SketchUp's views are available as Styles. These can contain custom lines, backgrounds, or watermarks. Many excellent styles have been premade and ship with SketchUp or are downloadable from the Internet. Styles can also be saved within animation tabs, which allows for multiple styles to be used in the same model.

Styles are applied from the Styles window—where you can also make customizations and edits (see **Figure 5.2**).

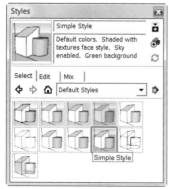

Figure 5.2: The Styles window

SketchUp Pro Includes LayOut

If you own SketchUp Pro, then you also have access to LayOut, a presentation and drawing preparation software that ships only with the Pro version of SketchUp (see **Figure 5.3**). LayOut provides sheet-based presentation options that are not available in the free version of SketchUp. For example, you can add a sheet border as well as a text field, and you can arrange multiple SketchUp views—complete with Styles—on the same sheet. This allows for the inclusion of SketchUp in the document creation process without the need for third-party software.

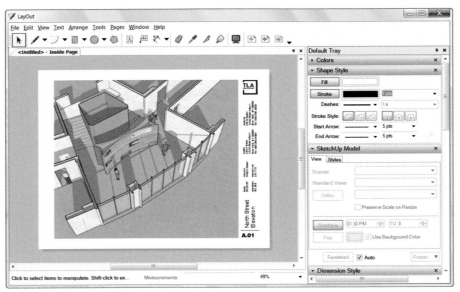

Figure 5.3: LayOut main screen

The Magic of Styles

Because Styles are central to SketchUp's own presentation options in the sense that they allow customization of the view as well as linework, let's review working with Styles with a simple example—albeit an uncommon one.

Example 5.1: Making a T-Shirt with SketchUp

1. Let's start with a simple model—a Tetris-like assembly of colored blocks. Of course, you may pick whatever design you like. Let's assume that we wanted to make a sketchy-looking T-shirt with this "logo," and we need to preview how it will look.

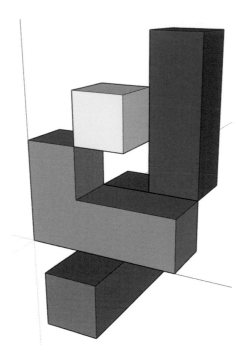

2. Open the Styles window and select the Edit tab. The icons at the top of the tab provide access to the various settings. After clicking on the first icon (the Edge settings), change the settings for Edges and Profiles as shown here.

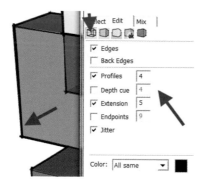

3. On the same tab, click on the fourth icon (the Watermark settings). After you click on the plus symbol to add a new watermark, select the background image that will appear behind our SketchUp model. Because this is a design sample for a T-shirt, I used a torso picture of a male figure in a white shirt onto which we will layer our design.

4. On the last icon on the tab (the Modeling settings), deselect the display of the model axes and any other distracting elements.

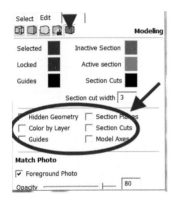

5. At this point, the graphic is overlaid on the image of a person. You can now zoom and pan the 3D model into position. The end result is shown here (left before zoom and right afterward).

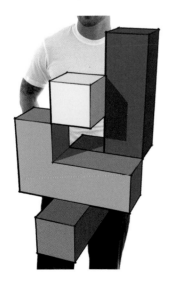
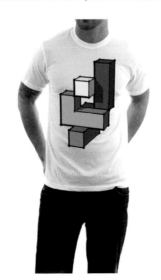

6. While this might be quite an uncommon use of Styles—let alone SketchUp—it showed you the settings you can modify in the Styles window. Feel free to save the style for future use (right-click on the style icon in the Styles window and select "Save As"). You can also now save the image (under **File → Export → 2D Graphic. . .**) for further processing in Photoshop or directly for selling your next T-shirt idea.

TIP	

If you actually want to print the graphic on a T-shirt, remove the watermark first and do a Zoom Extents so that you can save the graphic at a high resolution.

One point to remember: Edge and profile thickness will be scaled when you export at a higher resolution. Adjust those thicknesses until they look good at the desired resolution.

As you just saw, Styles in SketchUp let you customize:

- **Edge (line) appearance**—Edge/profile size, depth-cue, and jitter, but also whether you want to use a sketchy style.
- **Face appearance**—Visual styles such as hidden lines or transparency.
- **Background colors**—Colors for the background, sky, or ground.
- **Watermarks**—You can add an image as a watermark in front as well as in back of the modeling environment.
- **Visibility of nonmodeling items**—You can turn visibility on or off for items like axes and section cuts.

You can download new and interesting styles from several websites. One great resource is this page: www.sketchupartists.org/presentations/sketchup-styles.

PRO ONLY	

Style Builder

SketchUp Pro ships with a small program called Style Builder. This software allows you to use your own hand-drawn pencil strokes in a custom style—giving you the ultimate means to customize and personalize the visual appearance of the SketchUp-generated view.

What If I Want More?

Having reviewed SketchUp's built-in visualization tools, it becomes clear that one missing feature is photorealistic rendering. While shadows add a level of reality to lighting, the only light source accurately used to display shadows is the sun. Any artificial lights (such as streetlights or interior lighting) and any indirect lighting effects cannot be displayed.

Materials are similarly limited. SketchUp's materials—while giving a surface more realism—look flat and are unable to display reflectivity or accurate light transmission, let alone refraction (as would be the case with glass, gems, or water).

These shortcomings can be remedied only by using photorealistic rendering software. Fortunately for today's SketchUp user, there are many options available—from free programs to movie-quality (and expensive) software packages. The following sections present some of these and introduce you to common techniques used to create high-quality photorealistic renderings.

See **Figure 5.4** for a comparison of the display styles that are possible in SketchUp without a rendering software.

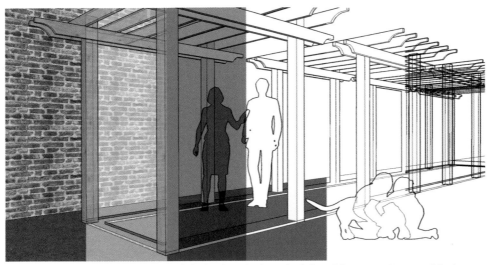

Figure 5.4: Comparison of different default SketchUp display styles (left to right: textured, x-ray, solid color, conceptual, hidden line, wireframe)

Overview of Rendering Methods

Photorealistic rendering in SketchUp (or any 3D modeling software, for that matter) uses a different approach from what we saw when we looked at SketchUp's own styles. In the case of styles, SketchUp uses your computer's graphics card to "render"—or calculate—the view appearance. In that process, the view and all visual features (such as custom linetypes) are being created pixel by pixel by your graphics card at a rate that is faster than your eye can perceive it being refreshed. As a reference, your monitor has a refresh rate of at least 60 hertz (60 screen updates per second).

Because SketchUp's visual styles can be displayed at roughly the same frame rate that your monitor uses, you can actually move around a model that has styles applied and view or edit it in real time (this performance depends on the model complexity and style complexity, as well as the graphics card and CPU quality, of course—the higher the demand on the graphics card, the slower it can process the view and the more jagged any motion becomes).

Photorealistic rendering, on the other hand, has the capability to accurately estimate material properties such as reflection as well as light properties such as colored lights or soft shadows. This, however, comes at a computational cost. In most rendering software, these computations are not performed by the graphics card but by the CPU (your computer's "computational heart"). Only the newest rendering software also processes some or all of these computations in the graphics card itself (commonly called the GPU). In any case, the necessary calculations typically are too slow to allow for real-time rendering at a high level of quality. Therefore, whenever you set up a view for rendering, you will send it from SketchUp to the rendering software and then wait for the rendering calculations to finish.

As you will see in the following sections, software is available that tightly integrates itself into SketchUp and that will execute all of these steps without requiring you to move files between different software. Some of the available programs are actually very fast (mainly because they now use both the CPU and the GPU), which will give you near-instant feedback on the rendered appearance once you apply a change (e.g., a material change or a camera move).

Before we look at rendering techniques, let's quickly review how rendering works.

What Is Photorealistic Rendering?

In real life, a light source like the sun or a desk lamp emits light particles (photons) in either a single direction, as is the case with a highly focused lamp or a laser, or radially in all directions, as with the sun or an unshielded light bulb. Briefly ignoring Einstein, we'll assume that, for all intents and purposes, these particles travel in a straight line.

These light particles then land on an object (your work desk or this book, for instance), where they are either absorbed by the surface (and turned into heat in the case of a black surface), reflected (as is the case with most surfaces), or refracted while they travel through the material (their direction is "bent," as is the case with water). Different materials exhibit different absorption, reflection, and refraction properties—water, a mirror, and a brick are three fundamentally different examples.

If a light particle is reflected, then it has the chance to land on another object—or many more, as happens in a hall of mirrors, for example. Depending on the surface properties of the materials these particles land on, reflection is either very directional (e.g., with a mirror) or scattered (e.g., on a painted wall). See **Figure 5.5**.

Figure 5.5: Various reflections in real materials (brushed metal, coated wood, natural stone, shiny metal)

Finally, as you likely have observed multiple times before, light has the tendency to become dimmer the farther away we are from its source. Imagine looking along a street with street lamps at night. This effect, called *attenuation,* means that light particles can't travel forever—they lose energy the farther they travel. (See **Figure 5.47** for an example.)

As you can see in this brief description, materials and particularly surface qualities like color, reflectivity, and refractivity have a large impact on how light travels and how our natural environment becomes visible to our eyes or a camera lens.

Photorealistic rendering software must be able to re-create these physical phenomena as accurately as possible. If we want our renderings to look like photos (or even better), then the software must be able to represent all of the aforementioned light and material properties as realistically as possible. However, it must be able to do that in a time frame that we as the users can live with. As you can imagine, any computational simulation of complex natural behaviors can be extremely resource-intensive. Therefore, rendering software is often designed to take some shortcuts that approximate the "correct" solution just to keep computation time low. It is a measure of good software that it is able to create high-quality renderings in a short amount of time.

So how does rendering software work? Without getting too deeply into the technical details, it is fair to say that it basically works like nature, in that it uses light particles that travel through space. Depending on the computational approach, however, it may be more efficient to send out these particles "in reverse" from the viewer's eye (i.e., the camera view), instead of from the light sources. Sometimes a combination of both techniques is used.

When these particles meet a surface, the software queries the surface material's properties and modifies the particle's color, direction, and intensity to reflect the material's absorptive, reflective, or refractive properties before sending the light particle back on its way. (See **Figure 5.6**.)

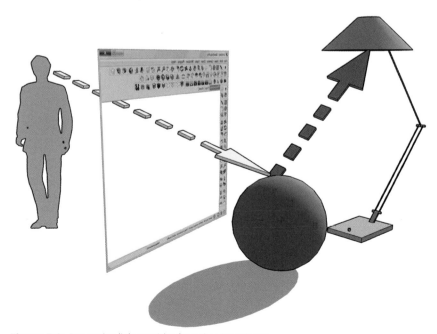

Figure 5.6: Ray-tracing light particles from the viewer's eye

It should be noted here that for computational efficiency, the number of "bounces" (or material interactions) is typically limited to a number that is low enough to reduce computation time, yet high enough to make the rendering look realistic. You can often experiment with this setting in your rendering software.

The following natural phenomena can be simulated by rendering software. Actual implementation and quality of the effects can vary among rendering software, though.

- **Material appearance**—This is commonly called *shading* and includes the interaction between light and a surface's material.

- **Texture**—This is the photographic appearance of the material. Often, seamlessly repeating images are used to represent a particular material (e.g., a brick pattern), but this could be accomplished by single images or a computational method as well (e.g., for solid colors or a patterned surface). SketchUp's materials are all seamlessly repeating *tiled* images. (See **Figure 5.7**.)

Figure 5.7: Various surface textures (rusty metal, wood, paper, canvas, solid colors)

- **Surface roughness simulation**—*Bump mapping* or *displacement mapping* can at least add the appearance of geometric detail and surface irregularities to an otherwise flat surface. Displacement mapping typically goes further than bump mapping in that the underlying geometry is modified. Bump mapping modifies only light reflection behavior. (See **Figure 5.8**.)

Figure 5.8: Material bumpiness: patterned rubber, natural stone, canvas

- **Reflection and shininess**—This adds fully or partially reflective properties to a surface. In rendering, this property is often adjusted using a *specular* texture or color.

- **Refraction/diffraction**—This describes how light direction changes when it moves from one material onto another.

- **Translucency**—This is the degree of opacity of a material.

- **Subsurface or particulate scattering**—When light travels through any material, particles (in the air, for example) reflect it and we can see effects like the "beams of light" that we notice in the forest when sunlight breaks through fog. A similar behavior occurs with a material that lets light travel into it, where the light then scatters and makes the material appear as if it is partly illuminated from inside—good examples are milk and alabaster. (See **Figure 5.9**.)

Figure 5.9: Materials with subsurface light scattering: milk, soap, candle

- **Indirect illumination**—When a lighter-colored surface becomes illuminated by a light source, it invariably casts a light reflection on adjoining surfaces. This effect is called *indirect illumination* or *global illumination,* and it is a must-have in rendering software.

- **Caustics**—When light travels through glass or water, the internal refraction makes caustics appear. We know these from light ripples on the bottom of a pool or as light focusing behind a lens or a glass object. (See **Figure 5.10**.)

- **Shadows**—In reality, shadows are never as crisp and sharp-edged as in SketchUp. Depending on the light source, shadows either have blurred edges or are completely blurry.

- **Camera properties such as focal length and aperture shape**—When we take an image with a real camera, depending on the aperture settings, part of the image may be blurred because it was "out of focus." If a light source is located in the blurry region, then its outline will have the shape of the aperture of the camera, which is typically polygonal. (See **Figure 5.11**.)

Figure 5.10: Caustics behind a glass of water

Figure 5.11: Depth-of-field effect from an open camera aperture

Both of these effects can usually be simulated by rendering software.

- **Motion blur**—Another camera effect is motion blur. When we take an image of a moving object, such as a car, at a slow shutter speed, the image becomes blurry in the direction of its motion.

- **Animation**—Instead of just creating a single rendered image, it is possible to render animations. These are commonly created as successive renderings of single images, which then get stitched together into video using editing software. Because video frame rate can be anywhere between 25 and 30 frames per second (for fluid motion), quite a large number of single images need to be rendered to create a video of usable length. Advanced effects in rendered animations are animated objects (such as water ripples or driving cars) and light animations (e.g., a morning-to-night shadow animation).

Rendering Techniques

A variety of rendering methods exist, and every manufacturer of rendering software has its own proprietary modifications to common techniques. As previously mentioned, the typical trade-off is between quality and speed. Software implementations of even the same methods can vary widely in those categories between manufacturers. I recommend, therefore, that you test any rendering software that interests you and find out if quality and speed are satisfactory for your needs.

It is also common for rendering software manufacturers not to offer too many different methods in one software package. This may be a good idea because too many (and too technical) choices may actually be counterproductive by being confusing to the user.

Common methods are:

- **Ray-tracing**—This is the basis of every rendering method. Light particles are traced as they travel through the model and get reflected or absorbed by surfaces. By itself, this method often does not include light bounces (indirect illumination) and only relies on direct illumination from the sun or artificial lights.

- **Global illumination**—Any good rendering software will include global illumination. This technique considers light bounces, which yield interesting effects such as illuminated shadows and colors "bleeding" (reflecting) from a colored wall onto an adjoining white surface.

- **Biased and unbiased methods**—Rendering always consists of computational simulation. As with any computational method, the more precise it is, the more resources and calculations it needs. Therefore, many rendering techniques are actually efficient approximations of the accurate solution, which by its very nature introduces a "bias" of one form or another. Only "unbiased" methods can accurately evaluate light behavior and therefore can be considered to produce "perfect" renderings. Unfortunately, those methods require considerably more computation than a biased method.

  Examples of unbiased methods are *path tracing* (PT) and metropolis light transport (MLT).

- **Ambient occlusion**—This rendering method approximates shadows based on how many light particles can potentially arrive at any point on the model. As a result, inside edges or occluded parts become darker and outside surfaces are lighter. This method often produces soft shadows reminiscent of what can be seen on overcast days.

Rendering Software

Adding rendering capabilities to SketchUp can be a surprisingly easy and/or cheap step (admittedly, it can be quite expensive, too). At this point, various software programs are available that accomplish this task. In general, these can be classified as follows:

- **SketchUp-integrated software**—This group of programs installs into SketchUp as a plugin. The programs generally are accessed using either the menu system or a custom toolbar that becomes available after installation.

When you start a rendering, your SketchUp model is processed in the background and a window with the rendering result typically shows up right within the SketchUp environment.

Materials and lights are often handled within SketchUp as well and commonly use SketchUp's materials, components (for lighting components), and sun/shadow settings directly.

TIP

If you use only SketchUp for rendering and like the convenience of having an easy and integrated workflow, look for integrated rendering programs. The only downside can be that these programs are designed to be so user-friendly that they might not allow much customization.

- **External programs with exporter plugin**—These are often stand-alone programs that can render scenes from a variety of 3D modeling packages (SketchUp, Rhino, Autodesk software, etc.).

  The user interface in SketchUp is similar to those of the first software group, because they integrate into SketchUp using their own exporter plugins. When you start the rendering, however, the 3D model data is exported and directly loaded into an external program, which renders the image.

  Depending on the rendering software, you may be able to modify any of the rendering settings (materials, lighting, etc.) even up to a very high degree in the external rendering software.

TIP

This approach is ideal if you use multiple 3D modeling software and would like to use the renderer with all of them. Because some of these programs offer extensive customization capabilities, they are often well suited for advanced users who look for more control.

- **External programs**—A large variety of rendering and 3D modeling/rendering software exists that is capable of loading 3D models using a generic file format (e.g., 3DS, OBJ, FBX, DAE, . . .). These can then be enhanced within the rendering software's environment by adding materials and lights.

  As long as the available import file formats are compatible with SketchUp's export formats, it is possible to use SketchUp's models with those programs. Some even use the SKP format as an input format.

  It is important to keep in mind that because they don't tie into SketchUp directly (and haven't been developed for it, either), rendering a model successfully often involves more work.

TIP

These programs are best for advanced users and those who seek custom features that only these programs offer. A benefit of external programs (which is therefore also true for the second software group) is that they usually can handle a large number of polygons and detailed textures quite well.

Depending on your preferences and your workflow requirements, you can choose between easy-to-use or highly customizable software. It is often a good idea to create a sample rendering using a trial version before you commit, especially to the more expensive programs. Other aspects to keep in mind when choosing are quality of help files and tutorials, availability of good materials, activeness of the user community (e.g., in forums), availability of plugins, and additional features such as animation support—if needed.

Multithreading and Network Rendering

In both of these techniques, a single rendered image is processed by multiple processor cores or even computers at the same time (i.e., in parallel). Given that today's workstations (and even notebooks) have multiple processor cores built in (as in a "quad-core CPU," for example), software can use each of these cores by sending a chunk (often a square portion) of the image to a different processor core in a *process* (or a computational *thread*). This can result in a significant increase in rendering speed. With enough computing power, near-real-time rendering is possible using this technique.

Almost all available rendering software can take advantage of *multithreading*. This means that if you have a computer with four processor cores, the image can be split up and rendered by all of those cores at the same time. As mentioned, modern rendering software can often use the graphics card's processors (the GPU) to process some of those calculations, increasing the number of computational cores even more.

Keep in mind: Because your operating system (and any other running software such as your e-mail program) uses processor cores, too, it may become necessary to limit the number of cores used for rendering so that the operating system doesn't "freeze up" during a rendering. If this is a problem for you, look into the rendering software's settings—often there is an option to limit the number of processes used for rendering.

If you are an advanced user or you are setting up a rendering solution for a larger company, then it may be a good idea to look into the network rendering capabilities of any of the software in question. *Network rendering* is a technique whereby a single rendered image is processed by multiple computers (and their respective processor cores) at a time. These computers can be located in your office (e.g., using computers that sit idle overnight) or in what is commonly called a *render farm,* which is a similar setup of networked computers, this time with dedicated rendering computers that are provided (for a fee) by third-party companies.

The following list is a sampling of currently available rendering software for SketchUp. Because all of these software packages work with the free version of SketchUp, it is not necessary to own the Pro version to create renderings. For an updated listing, go to this book's companion website and look for "Rendering Software."

SketchUp-integrated:

- IDX Renditioner—Win/Mac—**www.idx-design.com**

- Indigo—Win/Mac/Linux—**www.indigorenderer.com**

- IRender nXt—Win only—**www.renderplus.com**

- LightUp—Win/Mac—**www.light-up.co.uk**

- nXtRender—Win only—**www.renderplus.com**

- Podium—Win/Mac—**www.suplugins.com**

- Render[in]—Win/Mac—**www.renderin.com**

- Shaderlight—Win/Mac—**www.artvps.com**

- Twilight—Win only—**www.twilightrender.com**

- VRay—Win/Mac—**www.vray.com**

External programs with exporter plugin:

- 3D Paintbrush—Win only—**www.3dpaintbrush.com**

- Artlantis—Win/Mac—**www.artlantis.com**

- Kerkythea—Win/Mac/Linux—**www.kerkythea.net**

- LumenRT—Win/Mac—**www.lumenrt.com**

- Maxwell—Win/Mac/Linux—**www.maxwellrender.com**

- Thea—Win/Mac/Linux—**www.thearender.com**

Sample of external programs:

- Lightwave—**www.newtek.com/lightwave**

- Vue—**www.e-onsoftware.com**

These programs typically install easily from an installation file. In SketchUp, they usually create a new toolbar that can be docked anywhere on your screen and a menu entry. Many also add tools to the right-click menu (e.g., when you hover the mouse over a light object).

Renderings in this book were created using the Twilight (see **Figure 5.12**), Kerkythea, and Shaderlight (see **Figure 5.13**) rendering software.

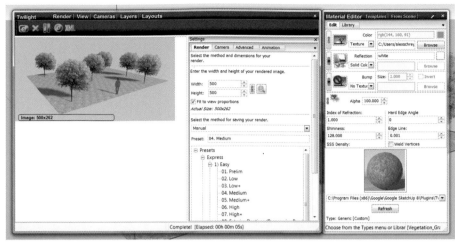

Figure 5.12: Software example: Twilight user interface

Figure 5.13: Software example: Shaderlight user interface

Set up Kerkythea and Create a Rendering

At this point you have read enough about the theory behind rendering and available software. It is time to actually get to work. To keep things simple (and cheap, because all the software we use in this short tutorial is free), I will use this chapter to give you a brief introduction to setting up the freely available rendering engine Kerkythea and doing some test renderings with it. This rendering software is available for both the Windows and Mac platforms. As you saw earlier, many other rendering software packages exist—all with their own strengths and weaknesses.

Please note: After this tutorial, instructions will be kept as software-independent as possible so that you can use them with whatever rendering software you end up getting.

Let's get started by setting up Kerkythea:

1. Download the Kerkythea installer from the official website: www.kerkythea.net/joomla/index .php?option=com_remository&Itemid=42&func=select&id=2 (Short link: http://goo.gl/5xyUa).

2. Download the SketchUp exporter plugin (SU2KT) and the SketchUp light components from the official website: www.kerkythea.net/joomla/index.php?option=com_remository&Itemid=42 &func=select&id=7 (Short link: http://goo.gl/S4dmh).

3. *Optional:* Download Kerkythea sample materials and models (trees, etc.) here: www.kerkythea.net/joomla/index.php?option=com_remository&Itemid=42&func=select&id=3 (Short link: http://goo.gl/MRO3y).

4. Install Kerkythea. The provided installer will install it like any other software. As you can see now, Kerkythea installs as a completely separate program (and can therefore be used for rendering 3D models even without SketchUp running). The only part that is contained in SketchUp is the exporter plugin.

5. Close SketchUp if you have it open. Then install the SketchUp exporter by copying and pasting the exporter files into the SketchUp plugin folder (usually C:\Program Files\ Google\Google SketchUp 8\Plugins on Windows and /Library/Application Support/Google SketchUp/SketchUp 8/Plugins/ on the Mac). *Important:* Make sure that you copy the plugin files so that the main Ruby file (su2kt.rb) is in SketchUp's main plugin folder and not in a subdirectory.

6. Install the light components into SketchUp's component folder by copying and pasting the files (usually C:\Program Files\Google\Google SketchUp 8\Components on Windows and /Library/Application Support/Google SketchUp/SketchUp 8/Components/ on the Mac).

7. *Optional:* Start Kerkythea and under the File menu, select "Install Library. . ." to install the material libraries by selecting the downloaded ZIP files.

8. Now start SketchUp. You should see two new items now: (1) An entry in the Plugins menu with the title "Kerkythea Exporter" and (2) a toolbar.

9. When you do the first rendering, the exporter will ask you for the location of the Kerkythea.exe file, so that it can open Kerkythea right after it exports a file for rendering, making the transition quite smooth. If for any reason, you are not able to set this up

correctly, don't panic! You can always save a scene for rendering using the exporter plugin (as an XML file) and later open it from within Kerkythea.

Once you have the software environment set up correctly, you are ready to give rendering a try. Create any 3D model you like—this tutorial uses an indoor scene where we can test some materials and lights. **Figure 5.14** shows the basic scene used for this example.

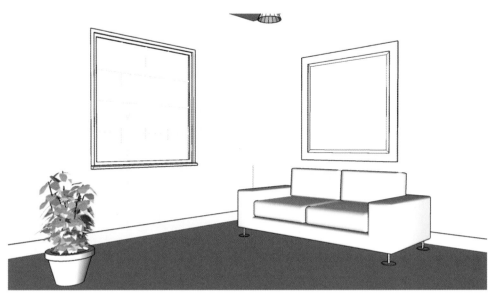

Figure 5.14: Room setup in SketchUp (box for walls and floor, sofa, window, and plant from the sample components, custom-modeled mirror)

Next, we need to apply some materials. Use SketchUp's built-in materials for this—add a wood material to the floor and any color material to the walls. Make sure all surfaces have a material applied and are not just painted with the default white color.

TIP

An important quirk to know about Kerkythea is the fact that the exporter groups geometry by material. The implication of this is that once you are in Kerkythea's environment and you wish to edit a material, clicking a particular face selects all faces that have the same material applied to them. Therefore, all faces that do not contain a material will be grouped because they actually contain SketchUp's default material.

As a good rule, make sure all faces in your model have some nondefault material applied.

Let's also look at the scene with the sunlight on. Go to the Shadows window and make sure shadows are enabled. Adjust object placement and sun position (i.e., modify time and day) to your liking. (See **Figure 5.15**.)

Figure 5.15: Room in SketchUp with materials applied and shadows turned on

To explore one of the most important features of rendering software—artificial lighting—we'll add a few lights to our scene. Let's first add two spotlights to the ceiling above the sofa. We'll use the exporter's built-in tool for this. Alternatively, you could place some of Kerkythea's light components into the scene, if you downloaded them when you were installing the exporter earlier.

Click on the Insert Spotlight icon on the SU2Kerkythea toolbar. The tool lets you pick a point where the spotlight will be located (click above the sofa on the ceiling). Then you need to pick a target point for the spotlight (just move down on the blue axis). Finally, a dialog opens with some light properties. You will also see a conical object appear in your scene, which represents the spotlight and its direction (it will not be shown in the final rendering). In the dialog, change light power to 5 but leave all other settings at the defaults.

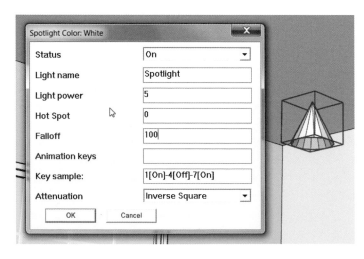

Now we have set up the scene properly for export to Kerkythea. All geometry has been created, materials have been applied, and all lights have been placed. As a final step, you could create several scene tabs in SketchUp, each showing a different view of our model. These would be exported to Kerkythea as differently named camera views.

To start the first rendering, click on the "Export model to Kerkythea" button on the SU2Kerkythea toolbar. This opens the export dialog, from which you need to select the options shown in this image.

Now simply save the XML file that the exporter creates in a convenient location.

If you look at the file location, you will notice that the Kerkythea exporter saved not only an XML file but also a folder whose name starts with "TX_." This folder contains all texture images for the exported scene. If you used SketchUp's texture positioning tools, then you will also notice that the exported textures have been "straightened" and exported exactly as they were visible on faces.

If you set up the link to Kerkythea in the SketchUp exporter correctly, then the 3D scene should open in Kerkythea right after exporting. If this doesn't happen, simply start Kerkythea and open the XML file from within Kerkythea instead. It should look like **Figure 5.16**.

Kerkythea's user interface is quite self-explanatory; nevertheless, take a minute to familiarize yourself with it. Here's what you see in the window:

Left of the model view is a tree outline of what is contained in the 3D space (models/objects, lights, and cameras). You can select something by double-clicking on the entry in this list, and you can modify parameters (e.g., light intensity or material properties) by right-clicking on it.

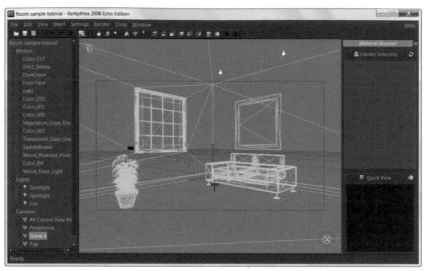

Figure 5.16: Room in Kerkythea after importing

Right of the model view is a material browser as well as a preview image of the rendered view. To view and apply materials, click on "Library Selection" and select one of the installed libraries. A double-click on a material applies it to the current selection.

Above the model view is a menu as well as a toolbar. On the toolbar you can find tools to change the view as well as start or stop a rendering. The menu offers many more tools, of which possibly the most important ones are under the Settings menu—you can control scene and light parameters as well as organize your material libraries there.

To be able to evaluate materials in Kerkythea's model view, press the "V" key to switch from wireframe to solid (or textured) view mode. This makes it easier to understand what is contained in your view.

Before we make any modifications to Kerkythea's settings, let's just start a test rendering. Select the **Render → Start** menu item or click on the round green runner icon in the top toolbar to do this. A dialog with various settings should open (the "Camera and Settings" dialog).

In this dialog, select the proper camera view that you want to render (choose "## Current View ##" for the perspective shown in the viewer). Also, set all other options to the ones in the preceding image.

TIP

It is usually a good idea to test a new rendering setup with low-quality and small image size settings. The higher the quality and the number of pixels in an image, the longer a rendering takes. Increase those settings only after an initial "draft" rendering gives you adequate results.

To determine final pixel dimensions, multiply the size at which you want to print the rendering (e.g., 5 × 7 inches) by the desired resolution (e.g., 300 dots per inch). This gives a minimum resolution of 1500 × 2100 pixels.

Your first rendering should now look like **Figure 5.17**.

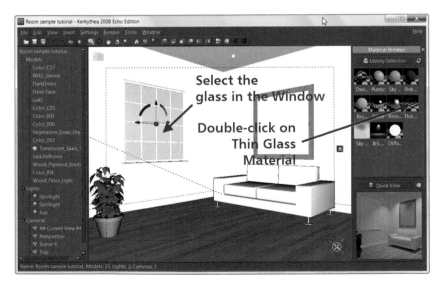

Figure 5.17: Kerkythea rendering with default parameters

As you can see, the window material didn't transmit any light. This can happen some-times. Let's go ahead and fix that by applying the Thin Glass material to the window glass. This step also shows you how to apply a material in Kerkythea.

Select the window glass to start this process. If you used a window like mine in your model, then you can select all of the polygons that make up the window by simply click-ing anywhere on it. Next, open the basic materials palette that came with Kerkythea in the Material Browser on the right side of the window. A double-click on the Thin Glass material will apply it to the selected window polygons.

The result after a repeated render should show the sunlight and its shadow on the sofa. As you can see in **Figure 5.18**, the mirror has a plain material and doesn't reflect at all. Let's fix this by applying a mirror material to it. This time, we'll make a custom material.

Figure 5.18: Rendering with window glass and sunlight visible

To do this, select the mirror polygons in your model first. You will then see the material highlighted in the tree view on the left side of the screen. Right-click on it and select "Edit Material . . ." from the context menu. A Material Editor dialog opens, where you need to make the changes shown here.

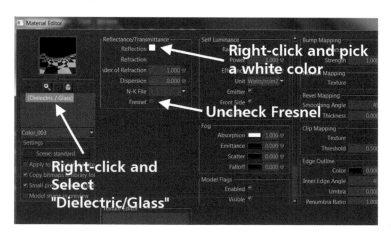

You won't see much of a change in the viewer, but if you redo the rendering, as shown in **Figure 5.19**, you'll see that the material now reflects like a perfect mirror.

Figure 5.19: Rendering with reflective materials

It might be a good idea to apply the same material to the sofa's legs, because they are likely made of polished chrome.

Let's modify some other settings. In the next step, let's give the sun's shadows some fuzziness—the crisp outlines of the window's shadows are not realistic. To do this, find the Sun item under Lights in the tree view on the left side of the screen. Right-click on it and in the Scene Settings dialog, modify parameters as shown here.

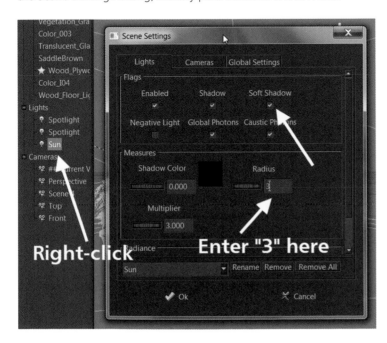

Test these settings with another trial rendering. Your room should now look like **Figure 5.20**.

Figure 5.20: Rendering with fuzzy shadows

At this point, you have set up a quite usable rendering, and it is time to do a final render-ing. Start a new rendering and select "08. PhotonMap—Fine + AA 0.3" as the setting. When it finishes, notice that the blurry shadows look much better than they did using the trial settings. Furthermore, you most likely noticed that it took much longer to create the better-quality rendering.

To familiarize yourself with the various rendering methods that Kerkythea offers, try ren-dering the same scene with these settings:

- **02. Ray Tracing—High + AA 0.3**—This setting does not calculate indirect light and considers only the direct sun- and spotlight for illuminating the scene.

- **08. PhotonMap—Fine + AA 0.3**—PhotonMap is the technique that we have used for this tutorial so far. This method calculates indirect light and fuzzy shadows well.

- **02. Metropolis Light Transport (BPT)**—This method is unbiased, which means that it uses real light properties and does not just approximate them. You will see that indirect light and some materials look much better using this method—especially in complex scenes. Stop the rendering process when you are happy with the result.

- **22. Clay Render (PhotonMap—Medium)**—This rendering method ignores materials and applies a default white material to all surfaces.

- **Path Tracing**—This is an alternative method used mainly for exterior scenes. It doesn't work well in our case.

Once you have created the final rendering, keep the Rendered Image window open and adjust the Tone Map settings to your liking. Then click on Save and save the rendered image as a JPG, PNG, BMP, GIF, or HDR file.

This tutorial helps you understand how to use a rendering program—Kerkythea, in this case. For more information about Kerkythea than we can cover here, read the Getting Started tutorials that you can download from this website:

www.kerkythea.net/joomla/index.php?option=com_remository&Itemid=42&func=select&id=6 (Short link: http://goo.gl/IkRP0)

A more thorough reference is Robin de Jongh's *SketchUp 7.1 for Architectural Visualization* (Packt Publishing, 2010), which explains rendering with Kerkythea in more depth.

As this tutorial demonstrates, Kerkythea offers many modification options that are useful for advanced users. Nevertheless, they can be daunting for others who may prefer simplicity. Thus, I encourage you to peruse the list of software presented earlier to find the rendering software that is right for you.

Setting Up Rendering Components

The approaches and techniques presented in this chapter are applicable to any rendering software. Therefore, they are presented in a very general but usable form. Be aware, however, that you may have to adapt the concepts slightly to work with your software.

Use this chapter as a reference when you are working on photorealistic renderings.

Modeling

First, let's look at some modeling-related issues. You will need to consider much of what is discussed here in SketchUp while you are creating your model and before you send it to the rendering software.

Edges and Faces

A SketchUp model of any size is made up of many faces that are bounded by edges. When you look at a shape such as the sphere shown here and turn on the display of Hidden Geometry (in the **View** menu), you'll see that it is made up of several polygonal faces, which in turn are bounded by edge lines (four in this example).

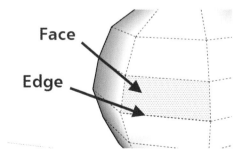

In addition, a model can have temporary guidelines (dashed lines) and guide points (little crosses) that you may have created when you were modeling using the Measure and Protractor tools.

When you send a model like this to rendering software, it typically uses only faces for rendering. Points and lines are assumed to have zero thickness and, therefore, cannot reflect any light, so they are generally ignored.

Two cases in which you might actually want to see lines in your renderings are (1) models that contain thin wires and (2) renderings that are intended to have outlines (either for more edge definition or for a more "cartoonish" look).

Let's evaluate this by working with a wire-supported tension-fabric structure model, as shown in **Figure 5.21**.

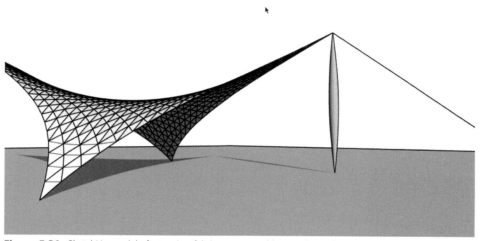

Figure 5.21: SketchUp model of a tension-fabric structure with guy wire and post

For the first case (making the wire visible), simply model it as a thin object to make it visible in the rendering. One consideration here is that it is likely very inefficient to do this with round objects (even though the wire may actually have a round cross-section). For example, the default circle in SketchUp has 24 side segments—extruding it along a straight line creates 26 faces (24 sides plus the top and bottom faces). Extruding it along a curved line creates many more faces because the geometry needs to fold around bends. (See **Figure 5.22**.)

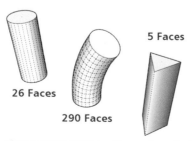

5 Faces

26 Faces

290 Faces

Figure 5.22: Number of faces after extrusion (left: default circle extruded along line; middle: default circle along arc; right: triangle along line)

Therefore, a good rule is to use the minimum number of faces when you model any object—especially when it is barely visible, such as in the case of a guy wire. To do this, you can preset the number of segments when you create a circle or a polygon by entering the number right after you start the tool.

The result of this approach can be seen in **Figure 5.23**, where the guy wire has been modeled using a triangle as a cross section.

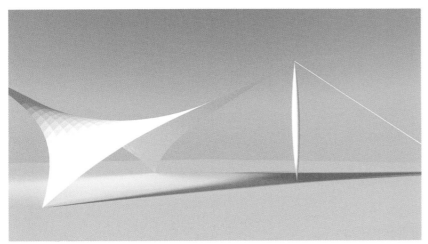

Figure 5.23: Rendering with guy wire modeled as extruded triangle

Now let's look at the second case: If you want to show the lines in the tension structure in your rendering, you have two options—the first one depends on your rendering software, however.

When you use your rendering software to assign a material to the faces that make up the tension fabric, look for a material feature called something like "Show Edge Lines." If this feature is available, the rendering software can actually render edges as a line (or even, in some cases, using a different material). If this feature is not available, a good approach is to overlay a rendered view with an edge-only view exported from SketchUp.

To do this, switch to hidden-line view in SketchUp (go to **View → Face Style → Hidden Line**) and turn off all shadows. Then export the view from within SketchUp. Before exporting, adjust any line properties in the Styles window to your liking.

Also, do a rendering of the same view (make sure the image size in pixels is exactly the same) and export it as well. You can now take these two images into Photoshop (or any other graphics software) and combine them using a layer arrangement similar to the image shown here.

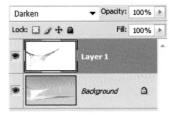

For this example, I also used a feathered white paintbrush to reduce the impact of the lines in some areas. The result of the combination now looks like **Figure 5.24**.

A further effect related to faces can be seen in **Figure 5.25**. Whenever you render a tessellated surface (like the fabric in this model), the polygons that make up the shape will render as flat shapes. As a result, the surface may not appear smooth in the rendering. This can often be solved with yet another material-related setting.

When you apply a material to the fabric in your rendering software, look for a feature called Hard Edge Angle. This may also be available in the general rendering settings. Similar to what you can set in SketchUp using the Soften Edges window, it is possible to pick an angle value below which adjoining faces will become smooth. For example, a value of 45 would mean that all neighboring surfaces that have an angle of 45 degrees between them will appear as one smooth surface. **Figure 5.25** shows an example of this method.

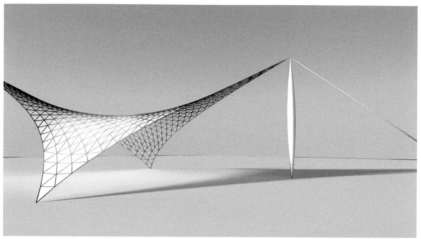

Figure 5.24: Combination image with rendering and hidden-line export

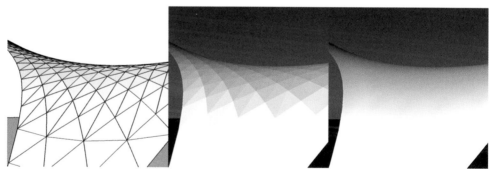

Figure 5.25: Smoothing of tessellated surfaces (left: SketchUp model; middle: no edge softening; right: full edge softening)

Of course, you can also smooth surfaces by increasing the number of polygons that make up the surface. You will have to do this using a plugin because SketchUp has no native function for this task. In addition, increasing the number of polygons may likewise increase render time, so having fewer polygons and visually smoothing them may be the most efficient approach.

Front Face and Back Face

SketchUp's faces have two sides to them: a front and a back. You can distinguish them easily by the default color with which each face is painted. The front is painted white, and the back is painted blue. Extruding a polygon and then removing some faces reveals the two colors quite nicely.

This has various implications for rendering. First of all, many 3D modeling and rendering programs display only the front side of faces—the back side is often simply not displayed. In addition, SketchUp has the ability to assign two different materials to the same face—one to the front and one to the back. This is also not always supported by other software.

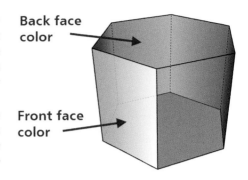

Back face color

Front face color

While many exporter plugins and integrated rendering software deal with this internally (where you don't have to think about it), it is important to know about this in case you need to troubleshoot. In any case, the best rule is to *always have fronts of faces point outward* (or toward the camera).

Fortunately, SketchUp has a built-in function that flips a face's orientation. To use it, select the faces you need to flip and then right-click on one of them. Select Reverse Faces from the context menu, and you will see the face color change. If you already have materials applied to those faces, switch the Face Style to Monochrome (in the **View** menu) beforehand. This suppresses all textures and lets you find reversed faces by showing you default face colors.

If you need to export a SketchUp model to use in any other 3D modeling or rendering application, look at the exporter options. Some of the options, accessible through the **File →** **Export → 3D Model. . .** menu item, allow for transfer of two-sided faces. (See **Figure 5.26**.)

Figure 5.26: Option in the DAE export dialog to use two-sided faces

Triangulation

Unlike many other 3D modeling software packages, SketchUp has the ability to use polygonal faces that have more than three edge points (or *vertices*). This allows us to use rectangular faces for walls—or even hexagonal faces (as shown in an earlier example).

While this is very useful for texturing—SketchUp's built-in texture positioning functions work on single faces only, for example—3D model export and processing by rendering applications typically leads to a triangulation of these polygonal faces.

This does not often cause problems, because the textured 3D model export into a rendering application is handled automatically by an exporter plugin. It is important to know this, however, for troubleshooting and for special cases.

One of these cases occurs when you use faces with a light-emitting material (e.g., to simulate a "light box" or a neon tube). While this can make a very nice-looking, diffuse light, it is important to keep in mind that the geometry of the faces that have the material applied should be as simple as possible—best is a triangle. If you use a circle with a light-emitting material, for example, the circle will be exported into the rendering application with many more faces than the one it has in SketchUp (see **Figure 5.27** for an illustration of this). The higher number of faces can increase rendering time significantly.

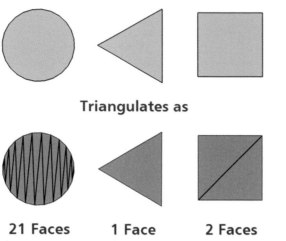

Figure 5.27: How similarly sized shapes triangulate upon exporting

Environment

Any of the objects that we render in SketchUp exist (if only virtually) in some environment. A building, for example, is situated in a landscape or urban environment, which at the ground level may have objects like trees, but above that has a sky (possibly filled with clouds) and a sun (as the obvious exterior light source). Another example is a small object, such as a statue, that would be rendered in a studio setting, with virtual artificial light sources such as point lights and spotlights. In the first case, the environment might be quite visible through windows, but in the second case, the background can intentionally be quite abstract and blurry. Either way, the background provides some context, scale, and possibly color balance to the rendered scene.

Another feature that can be accomplished by the environment is lighting. Background objects can reflect or diffuse light and influence the mood in the scene. Likewise, if an image is used for the background in a rendering, it can actually be the main or even the only source of lighting in the scene. This technique is called *image-based lighting* (IBL), and it is useful for creating realistic lighting scenarios—especially when high–dynamic range (HDR) images are used as a background.

Let's explore some of these effects.

Sky (Physical Sky, Sky Color, or Sky/Environment Image)

Any rendering software available for SketchUp usually is at least capable of rendering a physical sky. This is useful because SketchUp already has a physically correct daylight (and shadow) system built in, which can be enhanced by adding sky color, clouds, and haziness to the accurate sun shadows. The physical sky calculations ensure that atmospheric haze and sky darkening toward the zenith are reproduced properly.

Figure 5.28 shows the effect that a physical sky adds to a rendering. It still looks barren, but at least the lighting mood and the sky look more realistic and true to the time of day. Depending on your rendering software, the physical sky system may also add clouds and provide a parameter for "cloudiness."

Typical settings for a physical sky system are:

- **Turbidity**—The amount of haziness is set by this value. This affects brightness as well as light color based on the underlying algorithm.

- **Sun intensity and color**—Modify these to fine-tune the settings beyond what is built into the physical sky calculation.

Figure 5.29 shows the effect of varying the turbidity/haziness parameter.

Another option that may be appropriate for some renderings (especially studio-type setups) is to not have a sky at all and instead select a color (gray, for example) to provide the rendering with a uniform background. **Figure 5.30** illustrates this approach.

A third option is to have a full-screen image aligned behind the scene. Typically, this is possible within the rendering software by simply selecting an image from your hard disk. A drawback of this approach is that you must take care to match sun angles and light mood between the background image and the rendered scene before you use the image. Depending on the rendering software, these images also may not appear in reflections. Furthermore, this approach is not useful for animations where the camera or the sun is animated. A benefit, however, is that you can easily take a photograph of the environment of, for example, a new building from exactly the position where a window will be and then use that image as a realistic-looking backdrop for an interior rendering. (See **Figure 5.31**.)

A side benefit of using this approach is that later you can easily remove the sky (and replace it with a photograph) in Photoshop or any other graphics software by selecting and erasing the background color. This is especially true if you "green screen" the back with a bright green color (0,255,0 in RGB values).

Figure 5.28: Physical sky (top: late morning light; bottom: sunset light)—the sky color is based on the time of day

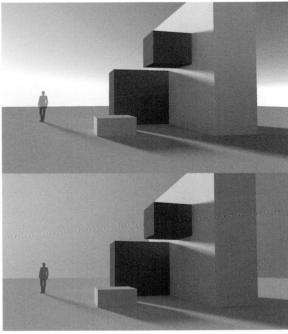

Figure 5.29: Two renderings of the same scene at the same time (top: low turbidity; bottom: high turbidity)

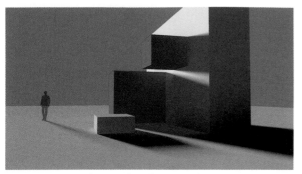

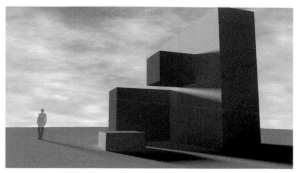

Figure 5.30: Rendering with uniform sky color

Figure 5.31: Rendering with background image of a sky stretched to fit the rendering size

TIP

It is a good idea to have a few neutral-color gradient images handy for use as backgrounds in these studio setups. For example, create an image with a gray gradient (from light gray to dark gray is usually best). Experiment with gradient directions, too. The accompanying images show some sample gradients that you could use. Always remember that a subtle gradient is usually better than a stark one—for example, a white-to-black gradient or a gradient between two primary colors.

If you intend to use this gradient as a "wraparound" spherical background (discussed later), make sure the gradient is oriented perfectly vertical.

IN-DEPTH

Backdrop Images in Your Model

If you need the ability to realistically move the view (or the camera) through the scene and have the background adjust accordingly (in what is called a *parallax shift*), a good approach is to set up your model as a studio scene with a backdrop. This usually gives you more control in the appearance of the background than if you applied a background image as an environment or sky image. **Figure 5.32** illustrates this setup.

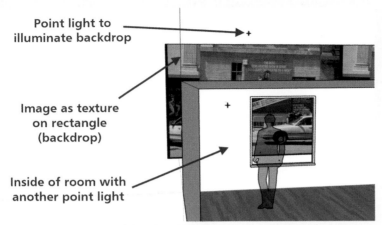

Point light to
illuminate backdrop

Image as texture
on rectangle
(backdrop)

Inside of room with
another point light

Figure 5.32: Studio setup with backdrop and lights

As you can see, I simply applied an exterior street scene photo to a vertically oriented rectangle (keep in mind that sometimes arched surfaces are better) that I placed a certain distance behind the window. Having this in place, I can then move the camera (the viewpoint) around within the room, and the background view shifts in a realistic manner. The result is shown in **Figure 5.33**.

In order to make this rendering work, though, I had to place additional light sources in the inside and the outside of this scene. Because the sunlight shines into the window, the backdrop would have been in the shade. Placing a point light above the room (and giving it a high light intensity) illuminated the backdrop and even gave it a bit of overexposure—an effect that would be expected if you took a photograph from the inside of a room toward the outside.

Alternatively, you can make the backdrop image a light-emitting material and give it enough light power to make it look realistic. These materials are covered in a later section.

Figure 5.33: Two rendered views from inside the room (note the background shift)

When creating renderings with difficult light situations (as in this case, where we show the inside of a room that is artificially lit as well as the sunlit exterior), it is a good idea not to create renderings that are "too perfect."

One approach is to look at photographs with similar light situations and evaluate the imperfections, then try to apply a subtle effect that is at least similar to the rendering. Don't overdo it, though, or your rendering will look fake.

Examples are:

- Overexposed exterior backgrounds for interior renderings

- Light blooming in interiors (overexposure of light sources)

- Vignetting (a fuzzy darkening toward the edges of the image)

- Lens flare (a flare within a photographic lens caused by a point light source with high intensity—e.g., the sun)

Typically, you can apply many of these in postprocessing in your image-editing software more effectively than in the rendering software itself. However, some rendering software lets you add these effects automatically.

Night Renderings

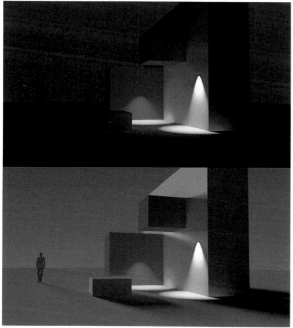

A common problem with sky renderings occurs when we want to render night scenes. In these scenes, we typically rely on artificial lights such as point or spotlights for illumination. However, when we use only those for our renderings, we generally find the result to be dull, badly illuminated, and uninteresting. You can see an example of this at the top of **Figure 5.34**.

The bottom of **Figure 5.34** illustrates a solution to this problem. Instead of using a solid background or sky color, this rendering uses a subtle gradient image. Furthermore, the scene is illuminated not just by the artificial lights but also by the sun—in this case, doubling as moonlight. After all, the night sky is seldom absolutely dark.

This approach gives the image more depth and general illumination. While this method isn't always appropriate, it is easy to execute. Just set the sun intensity very low in your rendering software and apply a bluish color to it. You may have to experiment a bit with the settings, but it is quite easy to create a moonlight effect using the sun.

Alternatively, you could go with an evening mood in your rendered image instead of a night shot. Try to create a sky that looks like the sun has just set, and start illuminating the scene from there.

Figure 5.34: Night rendering (top: only artificial lights and dark blue sky color; bottom: gradient background, light blue low-intensity sun to simulate moonlight)

For exterior night renderings, a good approach may be to create some color contrast between artificial lights and the night sky. To do this, apply a light yellow color to your artificial lights and a light blue color to your general illumination. This effect can be very subtle or quite dramatic. Adjust the settings to your liking, but keep in mind that subtle is usually better than overly dramatic.

360° Environments

Another approach to applying a sky and ground environment to a rendering is to use hemispherical skies, spherical skies, or a sky probe. The images in **Figure 5.35** illustrate what these look like. In essence, all of these are seamless images of the entire sky and ground as seen from a single vantage point. While a spherical sky image or a sky probe contains the entire 360-degree view, the hemispherical sky contains only the sky and omits the bottom half—the ground. The latter images are useful for oceanside renderings or any other setup where the visible ground is actually modeled in 3D.

Figure 5.35: Sky probe (top left); planar, spherical sky image (top right); hemispherical sky (bottom)

Using these environments has several benefits. First, the image provides a complete sky and ground environment, which can eliminate the need to add some background entourage (e.g., trees or a field) to the 3D scene. Another benefit is that if a spherical panorama has been created at exactly the location in which the virtual scene will be rendered (e.g., a location for a new house or a table in a studio where a statue is to be modeled), then the rendering will have accurate reflections and a correct background image for that location. Because the image "wraps around" the entire scene, it can also be positioned by rotating it around a vertical axis. Moving the camera view also looks realistic because the view of the sky background will change.

The most important benefit of using these environments, however, might be that the image can be used as the sole light source for the scene. This is called image-based lighting (IBL) and is discussed in the next section. For now, you can simply imagine that the entire 3D scene can receive lighting and color information from the image that was applied to the background. As a result, a scene that is illuminated in this way is lit in exactly the same mood (light color and general lighting ambiance) as the location where the environment image was created. See **Figure 5.36** for an example.

Figure 5.36: Scene with spherical sky image and image-based illumination

There are many places on the Internet where you can find spherical skies and sky probes. The following list of URLs is not at all exhaustive but should help you in your search for the perfect background image.

www.cgtextures.com—This site has several panoramic (hemispherical) skies

www.kerkythea.net/joomla/index.php?option=com_remository&Itemid=42&func=select&id=11 (Short link: http://goo.gl/9xlrw)—Various skies and sky probes from the Kerkythea website

http://ict.debevec.org/~debevec/Probes—Some HDR light probes

http://blenderartists.org/forum/showthread.php?24038-Free-high-res-skymaps-(Massive-07-update!) (Short link: http://goo.gl/r98W7)—Various sky probes and hemispherical sky maps

---

| **TIP** | You can very easily create sky probes yourself as long as you have a good-quality digital camera that has high image resolution. All you need is a perfectly reflecting chrome sphere. You can get one for less than $20 at some material suppliers—the one shown here is a precision-ground steel-alloy ball from McMaster-Carr (**www.mcmaster.com**). |

To create a sky probe, clean the sphere meticulously and place it in the desired environment. Then position your camera on a tripod and align it horizontally. Take a picture (RAW image format is best) of the sphere with a telephoto zoom and at a high aperture setting. Finally, crop and scale the image in your editing software so that the sphere completely fills a perfectly square image (as shown in the top-left image in **Figure 5.35**). If you loaded the image as a RAW file, export it as an HDR image; otherwise, just save it in the format the image came in. You can now use that image as a sky probe in your rendering software.

Ground

A final paragraph in this section should be dedicated to the ground. Whenever your scene includes a ground (terrain, an ocean, etc.), it is accomplished using any or a combination of the following approaches:

- **Model the ground**—At least the ground next to the viewer should be modeled in SketchUp and textured, which gives it a good level of detail. If you have something akin to an "infinite ground" (e.g., a view out toward the ocean), you will need to create a large plane for your ground and texture it.

- **Infinite ground**—Some rendering packages allow you to add to your scene an infinite ground at some height level. This is most appropriate for seascapes, where the water plane can easily be modeled this way. SketchUp by itself does not support infinite planes.

- **Sky environment**—If the sky environment (as discussed previously) includes a ground (e.g., in a spherical sky image), it may be sufficient to use the ground from this image as long as the image quality is detailed enough.

Figure 5.37 shows a sample scene of an oceanscape in which the water plane was modeled with a simple (albeit large) rectangle. The water ripples were added as a bump map, a technique that is discussed in a later section of this chapter.

Figure 5.37: Ocean scene with water plane

Lighting

Lighting is arguably the most exciting part of creating photorealistic renderings. This is where you can create spaces and moods that can't be visualized with the same level of realism within SketchUp. You can be very creative with light and approach your rendered scene like a theater stage where *you* are in charge of placing interesting lights and illuminating the space in various ways. The following sections review some of the common issues encountered when executing lighting. Feel free to combine any of the following methods to get the best result.

Ambient Lighting and Image-Based Lighting

In the absence of any direct light source, such as the sun or point lights, the scene must be lighted by environmental illumination. Depending on your rendering package, this can be done in one of two ways, both of which are global illumination (GI) techniques that rely on the rendering software's capability to calculate light scattering.

The first method is general illumination caused by a uniformly colored light that is emitted from an imaginary sky dome (a sphere that fully encompasses your model). Typically, you can simply select a sky color and then use any rendering method provided with your software. Because the result will look very much like an ambient occlusion (AO) rendering, that rendering method can give sufficient results. See **Figure 5.38** for an example.

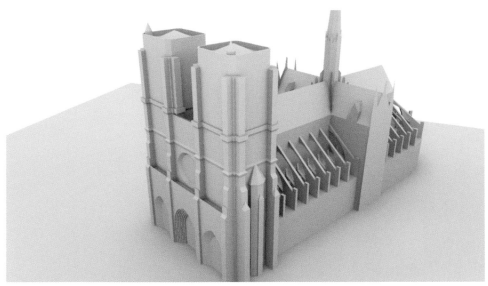

Figure 5.38: Ambient occlusion "clay" rendering of Notre Dame

This method is often used for "clay" renderings (like the one in **Figure 5.38**) where the material has been painted with either the default material or a neutral-color material (such as light gray). Often, this is not even necessary—the rendering software simply ignores all textures during rendering if Clay Render is selected. A benefit of this rendering method is that it allows you to evaluate shapes without being distracted by color and other surface textures.

A second method—which works well for clay renderings but especially for textured models—is image-based lighting (IBL). In this method, an image is applied to the imaginary "sky dome" similar to what was described in the "Environments" section. If this image contains a spherical projection of the sky and ground, it can be used to represent a full 360-degree background. Even if the image will not be visible in the final rendering, its color information will be used to apply environmental lighting to the rendered scene. The three images in **Figure 5.39** are examples of IBL with a high-dynamic-range (HDR) image as a light source. Note the different shadows and overall light mood and color.

This method is especially powerful when HDR images are used for the environment map. Because an HDR image stores color information at a much higher level of precision (32 bits) than a standard *low-dynamic-range* (LDR) image (e.g., 8 bits for a standard JPG image), lighting in the HDR renderings is very natural-looking. This approach also permits reuse of a particular interesting light setting (e.g., the environment in a church), whereby you can apply it to a completely different scene.

Figure 5.39: The same scene illuminated with different HDR environments (you can see them in the sphere's reflection)

It is often a good idea to combine image-based general lighting with in-scene lighting such as spotlights or even the sun. Experiment with your scene to find the best combination of lighting approaches.

You can find a large number of reasonably good quality HDR skies on the Internet. Follow some of the links that were presented in the "Environments" section to download some, and try this method for yourself.

TIP

If you are rendering an interior scene where all the ambient light is entering the scene from the outside through a window, one feature you want to look for in your rendering application is *sky portals*. A sky portal is a material that you apply to a window, where it basically functions as a light collector for the ambient light that exists outside of your model.

Using a light portal can yield a more even interior light and lead to better rendering times.

Sun

A scene that is illuminated by the sun primarily allows you to evaluate the interplay of shadows and geometry at any given day and time and at any location on the earth. Set these visuals in SketchUp's Shadows window. You can use them to easily create shadow studies that evaluate a building's overhangs and the reflecting properties of materials to add to the general illumination of a room, for example. It can often be useful to have the ability to create a rendered animation of sunlight and shadows over the course of a day. (See **Figure 5.40**.)

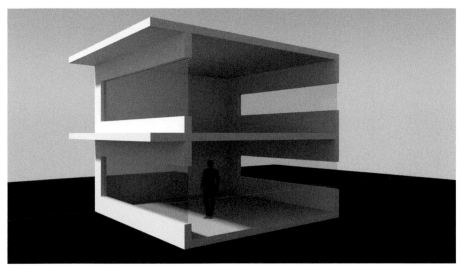

Figure 5.40: Solar studies on shading devices and windows using sun illumination and physical sky

TIP

If you own SketchUp Pro, you can adjust the north direction in your model by using the tools on the Solar North toolbar.

Use either the graphical (second button from the left) or the numerical (third button) method to adjust the north direction in your model.

While this is obviously useful for architectural visualization, using the sun can be practical for many other uses, too—for example, to simply bring out detail in a model when rendering it. The sun is also a convenient default light source over which you have full control simply by changing the time and day sliders in SketchUp's Shadows window. **Figure 5.41** illustrates a case where the flat sun angle was used to illuminate the model and bring out the shadows on the front of the Notre Dame building model.

When creating renderings using the sun, it is important to remember that the real sun's shadows are not as crisply outlined as they are in SketchUp. It is therefore important to give them a realistic "fuzziness" around the edges (see **Figure 5.42**).

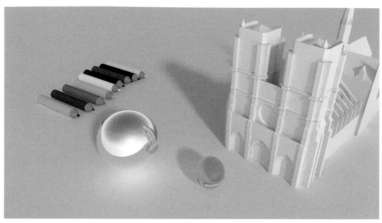

Figure 5.41: Sun illumination to enhance facade detail

Figure 5.42: The sun's shadows: (1) close shadows are sharp; (2) distant shadows are soft; (3) shaded areas in shadows are also soft.

Rendering software is usually capable of doing this. The only drawback is a potentially longer render time. If soft shadows are not enabled by default, look in your rendering software for settings related to the sun's radius (or something similar)—a larger-diameter light source produces softer shadows.

Omnidirectional Lights

Among the artificial light sources, omnidirectional lights—often called *omni lights*—are usually the simplest ones to work with. As long as the light is placed correctly in 3D space, all the information that is needed for it to work is its light intensity and color.

An omnidirectional light can be imagined as a point light that emits light particles from its location with equal intensity in all directions. A common use for these lights is therefore as lightbulbs in lamp fixtures. Another use case can be as a fill light in a dark area of a rendering; if you have an area that is quite dark given the light sources used, you can add a low-intensity omni light in that area (taking care to avoid unnatural shadows) to enhance detail.

Being a point light, the omni light by default creates hard and precise shadows. This may not be desirable in a particular rendering because natural light produces soft shadows. Therefore, rendering software usually includes the ability to create soft shadows by representing the point light as a sphere with a given diameter.

In some situations you may not want omni lights to produce shadows at all—for example, when using them as fill lights. In such cases, some rendering software adds the capability to turn off shadows on a light-by-light basis.

It is advantageous to use a studio-type setting to evaluate light properties. Therefore, in our present discussion we mainly use the studio setup shown in **Figure 5.43** for rigging lights. As you can see, it features one of SketchUp's default dummy 3D people as a model and has a continuous back and floor "canvas," as well as reflective sides. The lines visible in the setup have been divided into equal parts to provide points to which we can attach lights (simply by snapping them to endpoints of the line segments). As discussed, lines are not rendered and can therefore remain in the SketchUp model.

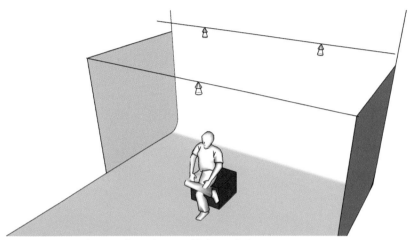

Figure 5.43: Studio setup for evaluation of light rendering

First, let's evaluate a default lighting setup consisting of three omni lights. **Figure 5.44** shows the result (on the left) and the setup for this (on the right). Two lights were placed above and slightly behind the model and one above and slightly in front of the model. This case used a white light color and the same light intensity for all three omni lights.

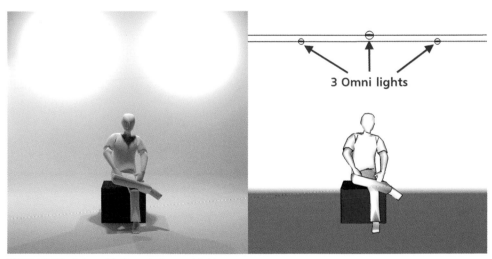

Figure 5.44: Setup with three similar omni lights

As you can see, the model is well lit with shadows that are not too dark. However, two problems are apparent.

First, the two rear lights are too bright. Their proximity to the background creates two overly bright areas. You might be tempted to think that what you see there are the actual lights. After all, if we take a photograph of a lightbulb, we get a similar result—a very over-exposed area around the lightbulb. That is not the case here. None of the standard lights that we use in rendering (omni lights, spotlights, IES lights) show up on renderings because they have no physical representation; they are merely computational light sources.

The second problem is that the front of the person is too much in the shade. Although this gives the scene a dramatic look, such an effect might not be intended.

Let's start by fixing the first problem. **Figure 5.45** shows a modified version of the same setup, in which light intensity of the background lights has been reduced drastically (to about 1/10 the intensity of the front light). These lights also received a blue light color to give the background more depth and interest.

Figure 5.45: Modified setup (reduced light intensity and added light color)

We can now go a step further and add one more omni light—this time, directly in front of the person (see **Figure 5.46**). This light functions similar to a flash that would be mounted on the observer's camera and is needed solely to brighten the shadows on the model. Because such a light would create a shadow on the background, shadows were disabled for this light.

Ideally a "stage" setup like this should have omni lights for the background and a spot-light for highlighting the person in the foreground. We discuss spotlights in a later section. First, however, let's look at an important light property: attenuation.

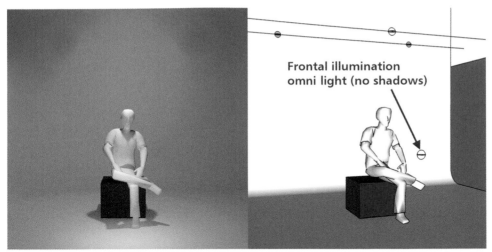

**Frontal illumination
omni light (no shadows)**

Figure 5.46: Setup with added light source in front of model

Attenuation

Any light (except IES lights) that you add to a 3D scene has a property called *attenuation,* which relates to the way light particles lose intensity the farther they travel from the light source. In reality, you likely have already observed this phenomenon. Just imagine looking down a street at night—the streetlights closest to you usually look brighter than the ones farther away.

Figure 5.47 illustrates this phenomenon and the various settings, using three spotlights. The top spotlight in the figure shows no light attenuation at all. As a result, the spotlight is as bright on the right as it is on the left—the only reduction in intensity comes from a "flaring" of the light.

The bottom two spotlights exhibit light attenuation—the middle one uses a linear model and the one on the bottom a squared one. These two look more realistic simply because of this property. It is therefore usually a good idea to use light attenuation for your light sources. If you want your rendering to closely resemble reality, opt for the inverse squared method.

Figure 5.47: Light attenuation in spotlights (top: none;
middle: inverse linear; bottom: inverse squared)

Spotlights

The second light type that any rendering software is capable of handling is spotlights. As the name implies, these shine light from the source in one direction only, producing a cone of light that can be used to illuminate details or provide dramatic light scenarios. Real-world equivalents are recessed lights (*potlights*) in a house and spotlights on a stage.

To evaluate spotlights, let's apply some to the same scene that we used in **Figure 5.46**. **Figure 5.48** shows the result on the left and the setup on the right. The only difference is the replacement of the three omni lights in **Figure 5.46** with spotlights, all pointing down, in **Figure 5.48**.

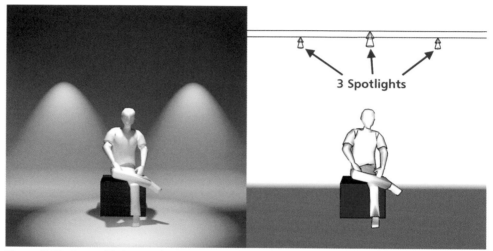

Figure 5.48: Setup with three similar spotlights

This scene already looks good and realistic even without any modifications. The two spotlights in the back are close to the background and therefore produce nice-looking light cones on the wall. All three have soft edges and produce soft shadows.

As we did with the omni lights, we can now modify some of the properties of the spotlights. For example, **Figure 5.49** shows what happens when we change the light color of the two in the back to blue. Together with the spotlight in the front, this immediately produces very nice stage lighting.

It is usually desirable that spotlights have soft edges. This is likewise rooted in reality because light cones, even from the most focused stage spots, don't have perfectly sharp edges. Nevertheless, let's explore what happens when we modify the related properties. **Figure 5.50** illustrates what happens when we switch from a soft-edged spot (on the left) to a hard-edged one (on the right).

The two spotlight properties of importance here are typically called *hotspot* and *falloff*, illustrated in **Figure 5.51**. Hotspot typically is an angular measure of the central light cone that will have 100% light intensity. Falloff describes the outer cone (again, as a value describing the "opening" angle) where light reduces in intensity from 100% strength to zero (at its outermost edge).

Figure 5.49: The same setup with some light colors changed

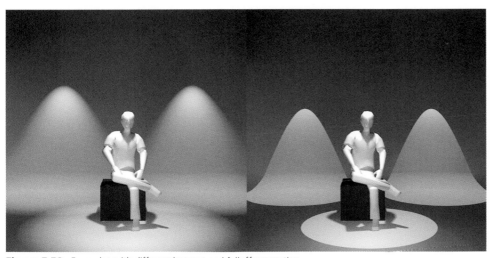

Figure 5.50: Examples with different hotspot and falloff properties

In **Figure 5.50**, the left image has a hotspot angle of 30 degrees, while the falloff angle is 60 degrees. This is what gives the spotlights the soft edges. The image on the right, in contrast, has both values set to the same amount (30 degrees). This is the other extreme and gives the spotlights perfectly hard edges.

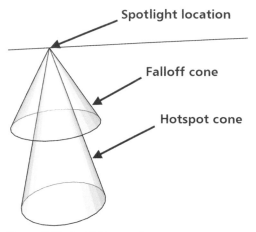

Spotlight location

Falloff cone

Hotspot cone

Figure 5.51: Spotlight properties

For your renderings, evaluate how a light would look in real life and adjust the values accordingly. As mentioned, always use at least a slightly larger number for falloff rather than making both numbers equal.

IES Lights

One method for adding realism to a light source without using either an omni light or a spotlight is to use Illuminating Engineering Society (IES) light definitions. As long as your rendering software supports this, using IES lights will immediately provide a high level of realism because the file that describes an IES light source has been generated by measuring a specific light fixture's light distribution. Using a specific luminaire's IES data (from an ERCO spotlight, for example) in your renderings gives you an accurate preview of how the actual light will look in a scene. (See **Figure 5.52**.)

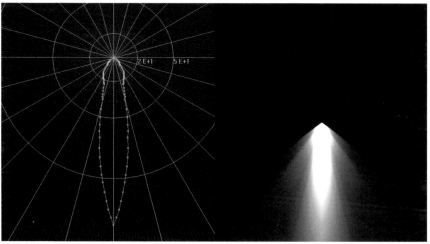

Figure 5.52: IES light data (left: data visualization; right: rendering—created with IES Generator 4)

Using IES lights is usually fairly simple. The first step is to get the appropriate IES file for a luminaire. Nowadays, manufacturers of high-end light fixtures usually allow you to download one file for each luminaire from their website. The following are a few websites where you can find IES tools and files:

www.rip3d.net/web/en.html#/free/downloads—This site provides a free IES file viewer and file creator software.

www.lightolier.com/home/file-finder.jsp?ftype=ies—One manufacturer's download page for IES files. Also browse other manufacturers' websites for their data (e.g., ERCO, Philips).

If you are curious about what IES light data looks like, open an IES file with a text editor. You will see several lines of text describing the light fixture and the manufacturer. Below that are several numbers that simply describe (in 2D) the light distribution—in other words, the location of the yellow points in **Figure 5.52**. Your rendering software can use this to reproduce the light distribution accurately in 3D space when you use IES lights in your scene.

Once you have the IES file you want to use, go into your rendering software and start applying it to a spotlight. Because the IES light data is directional (it can have uplight as well as downlight properties), you must make sure the orientation (rotation and tilt) of the light is appropriate. **Figure 5.53** shows how IES data is applied and visualized using the Twilight rendering software in SketchUp.

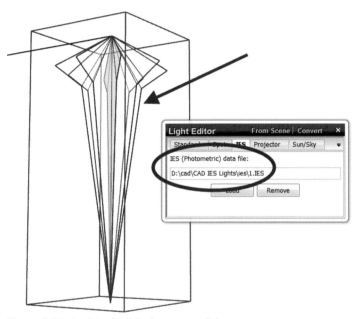

Figure 5.53: Applying IES light data to a spotlight

Once you render your scene, you should see the light distribution that was provided by the IES file (it should look less uniform than a standard spotlight). If the light intensity is not as needed, you will need to modify that property. Also, you can usually add a color to the IES light, if desired. It may be necessary to adjust light attenuation to even out light distribution as well.

Figure 5.54 shows the standard scene that we have been using throughout this chapter—this time, illuminated with IES lights. It is interesting to note that the IES data for the lights on the right indicates a vertical cutoff (it does not contain any data above the light source), which leads to the horizontal line between light and dark. I should also mention that if you applied IES data that describes an uplight (e.g., a wall sconce) to a downward-pointing spotlight, the spotlight will only emit light upward.

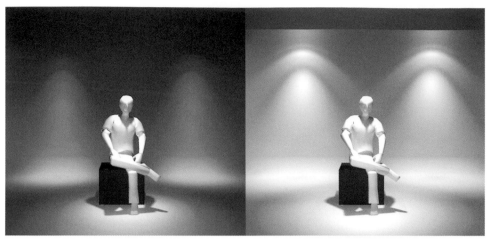

Figure 5.54: Our standard scene illuminated with two different IES lights

Recessed Lights

A common feature of contemporary buildings is recessed lights (sometimes called *potlights*). These are often a convenient choice in architecture because they do not project below the ceiling and they create an attractive-looking light cone. The question for us is how to render them.

Assuming that we might want to accomplish this with a simple spotlight, we could just model the rim of the recessed light and then add a spotlight in its center. When we do this, we get a rendered result like the one shown in **Figure 5.55**. Although the light cone looks appropriate, the light source itself is not visible and the center of the recessed light is not illuminated. As discussed, we would likely not see the lamp, even in a photograph, due to overexposure, but the center of the light *can* should at least be illuminated and appear white.

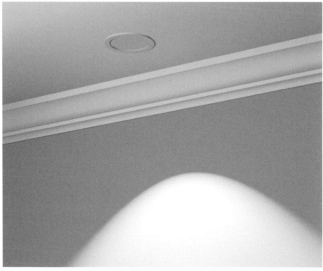

Figure 5.55: "Flat" recessed light without illuminated light source

There are a few ways to fix this problem. Depending on the situation you are dealing with, you might need to combine some of these. **Figure 5.56** illustrates three common approaches in SketchUp: spotlight, light-emitting face, and IES light.

Figure 5.56: Three methods to create recessed lights (left: spotlight; middle: light-emitting face; right: IES light)

The spotlight approach is similar to the one we used previously, with the main difference being that this time we modeled the recessed light as a true recessed *can*. This allows us to use the internal walls of the light fixture to capture light and therefore render the light source as an overexposed circle. The important property to change when using this approach is the falloff angle value, which must be large enough to include the side walls of the recessed fixture (e.g., 90 degrees).

The second approach, which uses a light-emitting face (a triangle, in this case), will be discussed in more detail in the next sections. Basically, it uses a SketchUp face and applies a light-emitting material to it. This approach has the benefit that we can control the size of the light source simply by modifying the size of the face. If you want to use this approach, make sure you use the simplest face possible. Remember that a triangle is exported to your rendering application as one face, while a circle has many more faces (which increases rendering time). It is also not as easy as with the other two approaches to tilt the spotlight (e.g., toward a picture on the wall).

The third method uses an IES light that contains a certain amount of horizontal light spread in addition to the downlight behavior (the same IES file used in the right image in **Figure 5.57**). The benefit of using this approach is that the IES light very likely reproduces the wall reflections of a true recessed light much better than either of the other two. **Figure 5.57** shows a comparison of all three approaches—note the double-arc light-beam projection with the IES light.

Figure 5.57: Rendered potlights from Figure 5.56

Use whichever approach is more applicable for your project. One thing to keep in mind, though, is rendering time. In the previous example, the spotlight produced the fastest rendering times by a factor of more than 2 (1 minute, 35 seconds) over the IES light (3 minutes, 15 seconds) and the light-emitting face (3 minute, 45 seconds).

Neon and Other Self-Emitting Light Sources

In addition to point lights and spotlights, there are many other light sources that we might need to render. One example that cannot be accomplished with the tools mentioned thus far is a simple neon tube, which is basically a linear light source. Another example is a 2' × 4' neon ceiling fixture, as installed in suspended ceilings; this is an area light source. Neither of these consists of just a point that emits light—it is the entire surface that (with varying intensity) emits light. As it turns out, these light sources are actually quite easy to set up—the only drawback is that rendering time can quickly increase if you are not careful with the settings. (See **Figure 5.58**.)

Figure 5.58: Neon light rendering

Rendering software usually enables this kind of setup by allowing you to apply a light-emitting material to a surface. Typically, you can just select one or more surfaces and apply the material to those surfaces as you would apply any other material. The software then assumes that every location on this surface emits light at a given intensity and without any single direction—light particles emit in all directions and produce a very soft, scattered light (similar to a light box).

Figure 5.59 shows an example of this approach. In this rendering, the model of a compact fluorescent lightbulb (CFL) was used as the sole light source in the scene. The faces that make up the light tubes were assigned a light-emitting material.

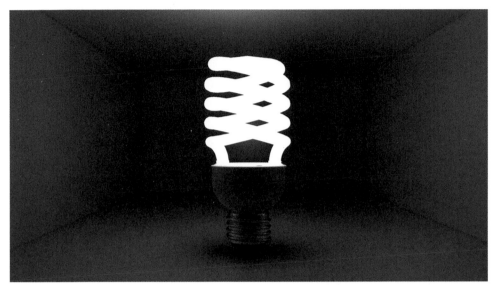

Figure 5.59: Model of a CF lightbulb with light-emitting material illumination

Although this approach is quite easy to implement and the resulting light is very even and scattered, keep in mind that rendering time is affected by using light-emitting materials. It is generally a good idea to keep to a minimum the triangle count of the faces to which this material is applied. Use this material preferably on triangular and rectangular faces, rather than circular or irregularly shaped faces.

When you use light-emitting surfaces, it is often possible to apply an image to the surface—some material properties, such as texture, also work for a light-emitting material. **Figure 5.60** uses an image of a plain 4' drop-ceiling light fixture as a texture on a single rectangular face to not only illuminate the scene but also give the light the appearance of two neon light tubes behind a frosted glass cover. This eliminated the need for actually placing light-emitting surfaces (representing the neon tubes) behind a frosted-glass material, which would likely have significantly increased rendering time.

Figure 5.60: Using an image texture as a light-emitting surface (left: rendering; right: photo of neon light fixture)

> **TIP**
>
> If your rendering software provides different rendering methods, it may be more efficient to render scenes with light-emitting materials using an unbiased renderer (e.g., MLT) rather than a photon-map GI renderer. This can vary with the software package, however.

Combining Lighting Scenarios (in Photo-Editing Software)

Various rendering packages nowadays have the ability to apply lighting scenarios once the rendering is completed. The benefit of such an approach is that after just one rendering it is possible to view a scene at different lighting stages—for example, with daylight, at night, or with different artificial light sources enabled. You can easily use this technique to create many different images from just one rendering.

Even if your rendering software does not provide this functionality, it is possible to simulate changing light scenarios in Photoshop (or any other image-editing software). The steps are as follows:

1. Render the same scene with different light groups enabled for each rendering. The left-hand image in **Figure 5.61** shows the three base renderings with (1) only daylight, (2) only the floor lights, and (3) only the downlights. It should be easy to disable lights individually or in groups in your rendering software. If that doesn't work, place all similar lights on their own layer and then enable only the appropriate layer before you do the renderings.

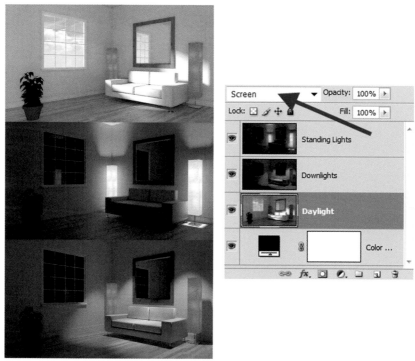

Figure 5.61: Three renderings with different lights are combined in Photoshop using "screen" layer blending

2. Load the rendered images into your photo-editing software's layers (**Figure 5.61** shows this for Photoshop).

3. Apply the "screen" layer blending mode to all layers. This "adds" light for each pixel when the individual layers become visible. You can also now adjust the opacity sliders for all layers to adjust the intensity or visibility of each of the light groups.

Figure 5.62 shows two resulting images that use a combination of the three base images from **Figure 5.61**.

Figure 5.62: Two images created using various combinations of the base images

Light Analysis in SketchUp

Especially if you are creating interior spaces in SketchUp, one common task in lighting design is figuring out light intensities (e.g., over workspaces). As part of this task, light amounts must be quantifiable (typically, in a unit called lux). Depending on the rendering software you are using, this may be a possible output that it can deliver. If the software is calculating light paths realistically anyway, why not use that data for analysis?

Figure 5.63 shows such an analysis. As you can see, light intensity has been colored in a false-color pattern, where the brightest areas are red and the darkest areas are blue.

To make such an analysis possible, you must use correct lighting data. Instead of using just light intensity factors for luminaires, make sure you enter correct wattages. Some rendering software also accepts watts per square area as a unit. This is especially useful for area lights such as light boxes and neon fixtures.

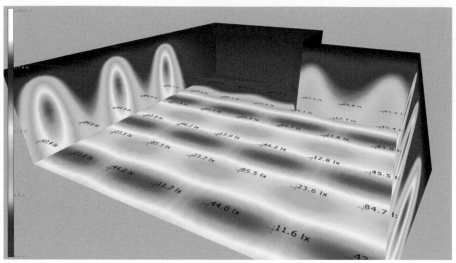

Figure 5.63: Light intensity analysis in the LightUp rendering software

Materials

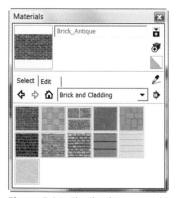

Figure 5.64: The SketchUp materials window

Much of the realism of a good rendering comes from using good-quality materials. In this section, we'll review various material types and how to set them up in your rendering software.

Fortunately for us, SketchUp already comes with a good array of materials—from simple colors to photorealistic repeating textures and even transparent materials. You can find them all in the Materials window (see **Figure 5.64**).

These materials work well in many situations. They are all seamlessly repeating textures, which allows their file size to be small and efficient. The main problems with them are related to close-up renderings, where it becomes obvious that the images have quite low resolution, and in some cases repeating patterns that make the textures look artificial. You can see both problems in **Figure 5.65**. In most other cases—especially when the view of the textured object is from a certain distance—the quality of the materials is sufficient for a good rendering.

Depending on the quality that your rendering is supposed to have, you will always need to choose among using SketchUp's materials, the materials your rendering software provides (if available), and making your own from scratch (e.g., from photographs). When you do this, you can vary these parameters:

- **Texture image quality**—A smaller texture image size often means more efficient rendering times (and a more responsive SketchUp display). For close-ups or large-size renderings, however, it is important that texture size is large enough to bring out detail.

- **Tiled versus single texture image**—Tiled, or repeating, textures (such as SketchUp's materials) are efficient because a small image can fill a large area simply by repeating itself.

As **Figure 5.65** illustrates, tiled textures can create repeating patterns that can appear unrealistic. It is, nevertheless, possible to use good-quality tiled textures (e.g., SketchUp's roofing or brick materials) that do not have the patterning problem.

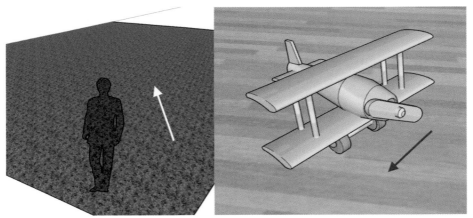

Figure 5.65: Problems with some SketchUp materials (left: repeating pattern; right: low texture resolution)

In some cases, it may be more appropriate to use a single texture image (that does not seamlessly repeat). A textured concrete wall that has graffiti on it would be such a case. For this, a single good-quality and reasonably straight image would be enough to provide an adequate texture as long as it fills the entire required area (e.g., the face of the wall).

- **Nontextured material properties**—Materials such as glass, reflective metals, a mirror, or a light-emitting surface do not need a textured image as long as an appropriate color is chosen for them and the rendering software deals with physical parameters such as translucency, reflectivity, light-emitting properties, and surface light scattering.

As demonstrated, it may be necessary to use an image-based texture along with these material properties to get the desired effect.

In general, it can be stated that adding reflectivity, translucency, and any other physical properties to a texture invariably increases rendering time. Therefore, if a material is farther away from the viewpoint or a material is not reflective at all, it is typically sufficient to use the unmodified SketchUp materials in your rendering software.

The images in **Figure 5.66** show a few materials that can be created by rendering software. They are (top, left to right): a default SketchUp color rendered as a matte material, a shiny plastic, a colored glass; (bottom, left to right): fully reflective mirror, a light-emitting material, and surface-textured water (notice the ripples).

Components of a Material

Let's look at the components that make up a material and discuss the various options related to them.

Depending on your rendering software, you have either very detailed or very little control over all these parameters. The upside of having rendering software that lets you control many material parameters is that you have ultimate flexibility and you can affect appearance in many creative ways. This often comes at the cost of convenience, though, and getting it right might involve more practice. On the other hand, the main benefit of having rendering software that gives you only the most important parameters is that the results are automatically of good quality. You might not have the option to tweak them, however.

Figure 5.66: Various rendered materials

When you shop around for rendering software, take a close look at the material editor and the options and check that the process and the quality work for you.

For each material that you use in your scene, evaluate the need for any of the following properties and use them as needed.

Texture or Color (Diffuse Texture)

The texture of a material is provided either by a color or by an image that is applied (*mapped*) to the surface of an object. By itself, the texture has no glossiness or translucency. Note, however, that in a global illumination (GI) rendering environment, the color of a texture can affect colors of surrounding materials by reflecting indirect light onto them—a good example of this can be seen in **Figure 5.28**.

You can apply a color to an object by simply painting it with one of SketchUp's color materials. Most rendering software uses SketchUp's materials as a foundation for adding features such as glossiness.

TIP

To apply a material to all connected faces in SketchUp, hold the Control key down when you apply the material to a face.

If you want to use a photo-based texture, then you can choose one of SketchUp's materials and paint it to the surface, or you can apply your own image to a surface. To do the latter, you have two options:

- **Insert the image as a texture.** Import the texture image using the dialog under **File →**
  Import. . . Then select "All Supported Image Types" in the File Type selector in the dialog and make sure "Use as texture" is selected on the right side of the dialog.

  You can then paste the image onto a face as a texture by clicking on its surface and dragging the image until its size looks correct (texture positioning will be covered later). Remember that the surface can't be in a group or a component. If that is the case, double-click it first to get into editing mode before you paste the image.

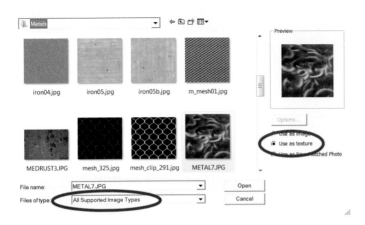

- **Make a new SketchUp material.** In the Materials window, click the plus sign to "Create a Material. . ." You can then load a texture for this material and apply it like any other.

  Make sure you adjust the size reference in the material editor to reflect the accurate dimensions. The next image illustrates this process.

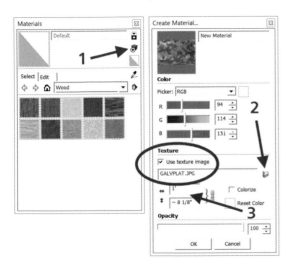

In **Figure 5.67**, you can see these methods in action. The tiles and the sand under the trees use materials that come with SketchUp. To give the concrete wall a rougher appearance than would be possible with SketchUp's tiled concrete material, a single image of a concrete wall was applied to its surface. This image does not tile well; therefore, it was necessary to position it so that it covers the entire wall surface.

Figure 5.67: Two texture types (single image and tiled)

TIP

You can modify any texture in SketchUp after it has been applied to a face using the **Edit Texture Image. . .** right-click context menu. This opens the image in the editor that you have preselected in SketchUp's preferences.

After editing and saving the texture image, the texture in SketchUp is automatically updated with the modified image.

Besides the textures that are contained in SketchUp's materials, there are many online resources where you can get good-quality imagery. Here are some examples:

www.cgtextures.com—An extensive resource with lots of good-quality (and even some tileable) textures.

www.2textured.com—A similar site.

If you search for "textures" or "cg textures" online, you will find a lot more. In addition to the texture images, software is available that can generate tileable textures based on a source image or a standard pattern (often complete with bump and other maps). A good example can be found at www.bricksntiles.com. Finally, there are many tutorials available online that show you how you can take an image and create a tileable texture simply using your image-editing software.

Reflection

When you render a SketchUp scene in your rendering software using default settings, you may notice that all materials look quite flat. They look very much like matte paint. In reality, however, most materials have some degree of shininess and maybe even reflectivity. Good examples are semigloss wall paint, the plastic or aluminum enclosure of your computer, magazine covers, and tabletops. Although their degree of shininess varies, most of them do have some. Only very rough materials (e.g., uncoated paper or soil) do not. (See **Figure 5.68**.)

Figure 5.68: Reflectivity of a walkway's stones

As you likely know, this is related to surface smoothness—the smoother the surface, the shinier a material is (up to perfect reflection). Polished metal and glass are materials that are very reflective—a mirror is a good example. Brushed aluminum (with its roughened surface) is quite reflective, but its surface scatters all light, thereby lacking a shiny and polished-looking reflection.

Figure 5.69 shows two materials that illustrate this. The ground material on the left has default properties and therefore looks quite flat. The material on the right has reflectivity similar to polished stone. As you can see, the reflection of the omni light (in the center of the image) is quite different from that of the material on the left. The light's reflection is visible on the ground, and there is not as much scattered light reflection close to the light source. Also, the checkered pattern on the back wall is reflected in a realistic fashion: Close to the wall, where the view angle is quite small, the pattern has a blurry reflection. Farther away from the wall and closer to the viewer (where the viewing angle is larger), the reflection fades out. This material appearance is typical of vinyl flooring or a semigloss tabletop, for example.

So what should you do to materials in your renderings? It is usually a good idea to give flooring, furniture, leather, and other partially or fully reflective materials at least a subtle reflectivity. This goes a long way toward realism in your renderings. Having said that, it is important to keep in mind that reflectivity invariably increases rendering time. Only perfect reflection renders fast, but that is obviously limited to mirrors and water or glass. Therefore, whenever a material is too far away from the viewpoint or nothing exists in the scene that could actually reflect in the material, it is usually okay to ignore reflectivity for that material. But this very much depends on your scene, and you need to evaluate this need on a case-by-case basis.

Figure 5.69: Comparison of flat and reflective floor materials

TIP	You can sometimes cut down on rendering time if you don't assign reflective properties to a material but rather use a subtle gradient for that material (which simulates a soft reflection). You must make sure it works with lighting and the general environment, however.

If your rendering software allows you to do this, consider adding a *specular map* to the material. This is similar to the bump map (discussed later) and is a grayscale image. When added to a shiny material, the grayscale values in the image determine the shininess distribution of your material (the darker it is, the shinier it will be).

Refraction

When you have transparent materials in your scene (such as glass, water, or gems), it can be important that these materials have realistic refraction properties. As mentioned, refraction refers to how light direction changes when it travels through a material. In reality, you likely have observed this—imagine holding a straight stick into a swimming pool, and notice how the lower part of the stick looks as though it is bent (see rendering in **Figure 5.70**). You can also see this behavior when you look at a thick glass or a prism from an angle.

Especially when your glass object has an observable thickness, it is important to include realistic refraction properties. This is the case when you model a glass vase or water in a swimming pool. Having said this, it is often not important to add refraction to architectural uses of glass. Windows can usually be modeled only with a transparent texture—as long as their reflection properties are accurate. Refractive glass would actually increase rendering time too much without adding much visual benefit.

Table 5.1 shows some refractive indices for various media. Use these as a guideline if your rendering software gives you the option to enter them for a material.

Figure 5.70: Refraction in water

Table 5.1: Refractive Indices

Material	Index of Refraction
Air	1.0
Water	1.33
Crown glass	1.52
Sapphire	1.77
Diamond	2.42

Figure 5.71 illustrates refraction in three prismatic transparent objects.

Figure 5.71: Refraction in three transparent materials (from left: glass, diamond, and frosted glass)

Bump

A *bump texture* or *bump map* is a very useful tool in a rendering software package's material editor; it can add significant realism at minimal cost to any material that has a surface roughness.

The best example to illustrate where we would use this might be a brick wall. In such a wall, the mortar is slightly recessed from the surface of the bricks. Depending on where the lighting comes from, we should see shadows in the mortar lines and highlights at the edges of the bricks. If such a wall were textured with a simple photograph of bricks, then changing light direction would not change the shadow appearance of the mortar lines, which would lead to a flat-looking wall. A bump map remedies this by giving the rendering software information where it needs to modify surface appearance so that the texture looks as though it is flat or recessed. (See **Figure 5.72**.)

Figure 5.72: The same texture rendered without bump (left) and with bump (right)

Other examples where bump maps are appropriate are bevel siding, flooring, paving, rough soil, and other textures where surface roughness creates gaps, creases, and depressions of less than approximately 1″ in depth. **Figure 5.73** shows a classic case where material bump should be used.

You can easily apply a bump map to any texture by providing a black-and-white image of the map and applying it in your rendering software's material editor (if the software supports this). It is important to remember that the map image has to have the same pixel dimensions as the texture image for both to align properly. During rendering, parts of the map that are dark will appear to recede while white parts will stay at the material's surface.

Figure 5.73: Brick walls—a classic case where bump is warranted

A common method that is often built in to rendering software is to use the desaturated (grayscale) inverse of the texture image as the bump map. This is a bit of a rough approximation that can work well, but it is usually a good idea to create higher-quality bump maps in image-editing software and then use two separate images for the texture and the bump map.

Some rendering software also offers a more advanced technique, called *normal mapping*. While bump mapping applies various levels of brightness to any location on a material depending on the depth information in the bump map, normal mapping modifies light properties based on a surface's "normal," or perpendicular, vector at any given point.

One method for creating a bump map in image-editing software is to start with a black-and-white version of the texture image and then invert it. Then use Brightness/Contrast adjustment to increase the black-and-white contrast in the image. Finally, use a white brush to remove any unwanted gray or black areas from the image. Don't forget to save the image at the same pixel size as the original texture. (See **Figure 5.74**.)

Figure 5.74: Two different bump maps (left: texture; middle: inverted black-and-white map; right: high-contrast black-and-white map created in image editor

Displacement

Depending on your rendering software, you may be able to add a displacement map to a texture as well. This technique goes one step further than bump mapping in that it doesn't simply modify light properties but modifies the surface of the object itself. Small or even larger details can be added to a surface through a displacement map—requiring less initial modeling and providing a higher level of realism because surface depressions or extrusions exist as actual geometry.

A good use for a displacement map would again be our brick wall—especially if larger bricks or round boulders were used for the surface texture. While bump maps fail to model wall edges properly (they don't show a mortar recess at the edge, for example), a displacement map can do this by creating geometry at this location. Other uses are terrain generation, whereby an entire three-dimensional terrain can be created from a black-and-white image (typically called a *heightfield*) in which pixel grayscale corresponds to altitude values.

Emittance

This material feature was already partly covered in the lighting section. What emittance can add to a material is its behavior of being a light source. This is most useful for aerial light sources such as neon lights and light boxes. It is also a good way to add lighting to display (or advertisement) boxes and television screens in your model. In those cases, it is also important that an appropriate texture is applied to the material, in addition to it being a light emitter.

You can see this in **Figure 5.75**. The "Stand Up!" text at the back wall is a neon-backlit sign that features the image of an evening sky as its texture. It was modeled using SketchUp's 3D text tool and by applying the light-emitting texture only to the front faces. Side faces and back faces of the text were left as is, which gives it the appearance (and shadows) of a display box.

In contrast, the light globes in the back corners are plain 3D spheres that have a white light-emitting texture painted on their entire surface.

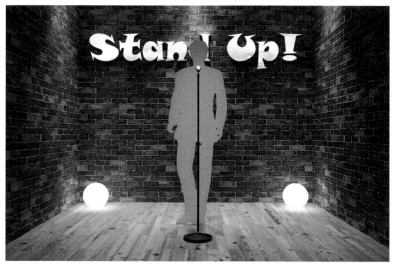

Figure 5.75: Emitting materials used for light globes and text-logo display

Some rendering software allows you to add a physical light property to a material, such as 100 W/m² illumination. Even if this is offered, it is often necessary to try out various settings until scene illumination looks just right.

> **TIP**
>
> Remember that the number of faces that have light-emitting materials applied to them directly influences rendering time. Try to keep those faces as simple and as few as possible.

Subsurface Scattering

Beyond materials that are opaque and fully reflect light and materials that are transparent and therefore let light pass through, there are some materials that let light filter through the surface of the material but then scatter it internally. The thicker the material, the less light travels through it. Good examples of these kinds of materials are wax, milk, alabaster stone, and translucent rubber.

To render these materials accurately, your rendering software must be able to handle a material feature called *subsurface scattering* (SSS). This material feature can then be assigned to any closed object (an object that has a volume and no openings in its surface). You can see the effect in **Figure 5.76**, where the candle wax material has subsurface scattering properties. As you can see, it is a white reflective material that internally also scatters some of the flame light, which was modeled as a combination of an omni light and a light-emitting flame object.

Figure 5.76: Three candles with subsurface-scattering wax materials

Using SSS for materials in your scene can put a strain on rendering time, and you should carefully evaluate whether you actually need it. For example, if you render a kitchen that has a few candles standing in a corner, you will likely get away with applying a simple white material (maybe with some glossiness) to the candles. However, if you render a close-up of a milk glass lit only by candles, you should use it for those two materials.

When you model translucent materials (as in the lampshade shown next), you have several options:

- **Use an SSS material**—This requires your lampshade to have an actual thickness and will likely result in a relatively long rendering time. Use the Joint Push/Pull plugin (see Chapter 4) to thicken the surface.

- **Use a partially transparent material**—This may work, especially if light can bounce inside the lampshade.

■ **Use fake transparency**—Add an emitting material to the outside of the lampshade to give it a luminous glow.

Putting These Features Together

Once you start working on materials for a rendered scene, it is usually a good idea to ignore the material features from the last few sections for the first test renders. Simply use the materials that you applied in SketchUp and work on lighting the scene. Then run a few low-quality test renders. Once you are happy with the view and the lighting, start refining materials and their properties.

TIP	When rendering, it often makes more sense to focus first on geometry, then on lighting, and on materials last. Often, however, it is a back-and-forth among all of these until you get the result you want.

When refining materials, keep your focus on the big picture. In other words, don't get lost refining materials that will barely be visible in the final rendering. That contributes only to the rendering time and not much to the result.

Follow these tips when you work on materials:

■ First, evaluate which objects in the scene need reflective (e.g., a mirror) or transparent (e.g., a window) properties. Since the scene will not look real without them, add those first. For reflective objects, start with a fully reflective material, and for transparent windows, use a simple transparent-reflective glass material that does not feature refraction or even dispersion—both of which render quite fast.

■ If you need any material-based lighting (e.g., from a neon tube), add light-emitting materials next. As mentioned, keep this as simple as possible to reduce rendering time.

- Next, look at large-area (in relation to the image size) objects, such as a wall or floor, and add appropriate properties such as reflection and bump. If you notice when doing a low-quality test render that adding these properties has no effect on the final result, undo especially the reflection properties, because your scene will render faster without them.

- Then focus on your key objects. Do you have a "showcase" object such as a kitchen countertop or a piece of furniture located close to the observer? Adjust its properties accordingly. If this object is made of glass, try a physically accurate glass material and see if its appearance improves. This will likely increase rendering time, but for the added realism it may be worth it.

- Finally, see if you need to manually adjust any of the material textures. In SketchUp, you can export a material's texture for editing in an image editor. This is useful if you need to add "edge dirt" (e.g., at the base of a wall) or fix tiling problems by making textures appear more random.

Tweaking Texture Positioning

One of SketchUp's best features is its ability to easily place textures from various sources and position them. As you have seen in this chapter, you can add textures using the Materials window or by importing images and placing them on faces as textures. Beyond these methods, you can also model using the photo-matching technique (which will not be covered in this book—if you are interested, you can find information in SketchUp's help system).

Once you have a texture for a particular object, it is important to get its positioning right. Fortunately, SketchUp offers tools to help with most common situations. Beyond those, we can use plugins. The following examples showcase techniques for this.

Example 5.2: A Wooden Tabletop

This example is relevant for any situation where a texture should extend from one face to other faces—be they in the same plane or not. We will use a positioned texture as reference and "paint" it onto other faces.

1. Let's assume that we need to place a texture of a slatted wooden board correctly on this 4' × 4' panel with a 2" thickness. To be able to place the texture, this object must not be grouped. If it were, we would have to double-click it and enter group (or component) editing mode. All surfaces have to be selectable.

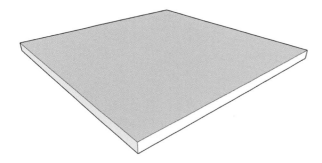

You can use SketchUp's texture positioning tools on textures that have been directly applied to faces. If you apply (or *paint*) a texture to a group of faces (or a component), all faces with the default texture will receive the new texture.

Remember: You cannot position texture applied to groups/components—even after double-clicking the group and getting into editing mode.

2. An easy way to get the initial positioning right is to import the image as a texture using the **File → Import...** dialog. Alternatively, you could create a new texture with the image and then adjust the positioning.

3. When placing the image on the surface, stretch it so that it fills the entire surface. After placement, SketchUp will simply "cut off" any excess (or tile the image). Because my texture image doesn't tile, I opted for filling the entire surface with the image.

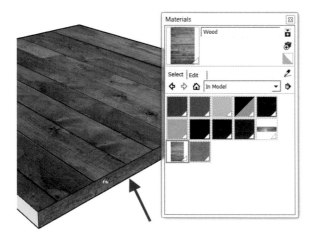

4. Since the texture has now been imported and a new material has been created, it is possible to use it from the Materials window by simply selecting it and painting any surface. As you can see in the accompanying image, in our case this has the drawback that its orientation is not correct—the wood grain runs perpendicular to the top surface's texture.

5. A better method here is to use the Sample Paint tool in the Materials window (the eyedropper) and sample the top surface's material before placing it on the side of the board. This method retains the orientation of the sampled material and then extends it to the painted faces, which gives our tabletop the appearance of properly aligned boards.

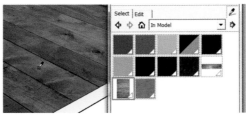

6. Although the texture already looks good, its positioning could be improved so that the board alignment looks more accurate. To do this, right-click on the texture you just placed and select **Texture → Position** from the context menu. This gives you SketchUp's alignment pins. As you can see in the image shown here, you now have pins to move, rotate, skew, and stretch the texture. For our purpose, just use the move pin (click and drag) to position the texture to your liking.

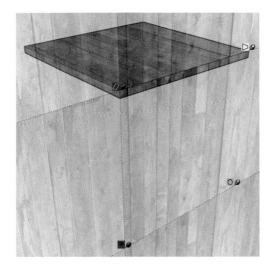

TIP

If you work a lot with textures in SketchUp (be it for rendering or geo-modeling), it is imperative that you become familiar with the pin positioning system. You can find more information about it in SketchUp's help system.

For now, keep in mind that you can click-drag to move, skew, or rotate the texture. If you want to position a texture relative to reference points, another method is to click and position (don't drag) the pins and then click-drag them to where you want the reference points to end up. This latter method takes some practice, however.

To reduce the visual impact of patterns that may appear when you use some of SketchUp's tileable textures, rotate them slightly. This is especially helpful when the texture tiling occurs along a straight edge in your model.

7. In my example (see accompanying image), I placed the side texture close to where the top texture ends so that color and seams line up better. Although this does not give us a perfect end grain for the boards, it provides us with something that looks similar to vertical-grain boards—especially if viewed from a distance.

These steps allow you to position most materials in SketchUp. In general, homogenic materials such as concrete, brick, and metals are easy to apply and straighten using the process outlined here. If you use non-homogenic materials (where one side looks much different than another side), it may be necessary to use other methods. **Figure 5.77** shows some of these approaches for the case of a rough-looking old wooden beam.

Figure 5.77: Three different texture-positioning methods in SketchUp (1st image: no end texture, 2nd: sample and paint; 3rd: paint projected texture; 4th: adding an additional end-grain texture)

The second image shows the same technique that we used previously: The top-surface texture was sampled and then painted on the end of the beam. As you can see, this method fails in this case because the end grain of the beam should be visible.

The third image shows an approach whereby the top-surface texture was first converted to a projected texture (right-click on the surface and select **Texture → Projected**). Then it was sampled and applied to the end. As you can see, this simply extrudes the texture at the end, giving it unnatural-looking stripes. Although this method may work well in other cases, it fails in this case.

The right image finally shows the only feasible method: adding a separate end-grain texture to the beam. This gives the wooden beam the most realistic look. Especially with wood (but also some other materials) it is important to keep a few "end" textures handy. Close-up renderings benefit greatly from those.

Example 5.3: Texturing Curved Surfaces

Curved surfaces have traditionally posed a problem for texturing in SketchUp (tiled textures, on the other hand, tile well on curves, too). Although it is possible to apply a continuous texture over a curved surface, positioning using the pin method previously described does not work on those surfaces (because they are made up of surface "facets").

TIP

If you need to align a tiled texture on a curved surface, turn on Hidden Geometry in SketchUp and align the texture on one of the faces. Then, sample and paint the aligned texture onto all other faces.

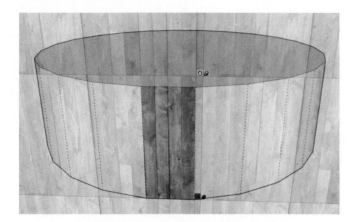

One method to deal with this is to add a planar object close to a curved object, place the texture on the planar object, position it there, and sample and paint the *projected* texture onto the curved object. **Figure 5.78** shows how a texture would look after this operation.

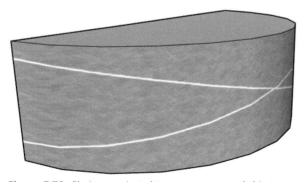

Figure 5.78: Placing a projected texture onto a curved object

As you can see, the texture looks good on the right side of the curve but stretches significantly on the left side. Therefore, this method works best for curved objects that feature only gentle slopes.

The steps outlined next describe how a texture can be precisely mapped onto a curved object using an unwrapping plugin. For this example, I will use Jim Folz's Unfold plugin (see Chapter 4) on a surface with a simple curve.

1. To begin, make sure hidden geometry is visible. We will need to see the dashed edges of all faces.

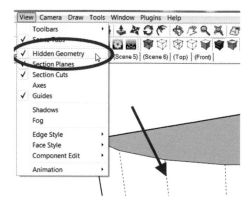

2. Select the curved part of the object and make a copy.

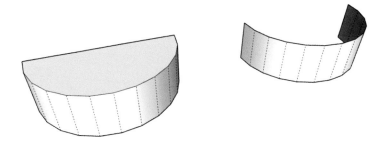

3. Now use the Unfold plugin to unwrap the curved surface into a planar surface. Of course, this is possible only with a surface that lends itself to unwrapping (commonly called a *developable* surface) such as a curved wall, a cylinder, or a cone.

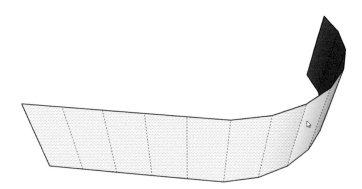

4. Use the eraser to remove all the dashed lines and create one continuous surface.

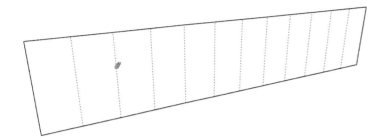

5. Now you can apply any texture image to this surface. Use the positioning tool to stretch and place the texture to your liking. For this example, I placed a cross-texture on the surface that illustrates its placement on the curve.

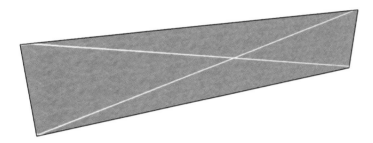

6. Right-click on the plane and select "Make unique texture." This creates a new texture that retains the current positioning and gets cropped to the face's edges of the surface. You can find it in the Materials window (under the In Model textures). Simply pick this one and apply it to the curved surface. To do this successfully, you must turn off viewing of hidden lines—the texture will then correctly be applied to the entire curved surface, as shown here.

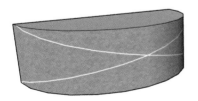 

Example 5.4: Texturing a Sphere

One object that cannot be textured correctly using any of the methods presented here is a sphere. A sphere in SketchUp is commonly produced by revolving an arc around an axis. As you can see in **Figure 5.79**, it therefore consists of rectangular faces of varying sizes and aspect ratios. If you use SketchUp's materials to texture it, you will likely get a result similar to the one shown in the middle image of **Figure 5.79**—a jumbled mess of texture orientations. You could manually position the texture of every single face, but this is not practical.

Figure 5.79: Texturing a sphere (left: sphere showing hidden faces; middle: default mapping in SketchUp; right: spherical mapping using UV Tools plugin

Fortunately for us, there is another plugin that can help out here. UV Tools by Whaat (Dale Martens) provides spherical and cylindrical mapping. All you need to do is select all faces on the sphere and right-click on one of them. The context menu then offers you both mapping options. If desired, you could now also adjust texture size in SketchUp's Materials window.

Objects

As you can imagine, a rendering of a kitchen looks quite sterile and uninhabited if there are no objects (e.g., a vase, a plate with fruit, a wine bottle) on the countertops. Likewise, an external rendering of a building looks artificial unless there are a certain number of realistic-looking plants and exterior objects in the scene. Including people in a rendering often lends the scene realism and scale and gives viewers an intuitive way of imagining themselves in the environment.

As you can see in these examples, objects and entourage (defined in the widest sense) are important for our renderings. So how do we include them?

In general, there are two methods, each with its respective advantages:

- **Include in the rendering process**—This involves getting 3D or 2D models into SketchUp (or your rendering software) and placing them in the scene. While this seems to be the most intuitive method, it might be labor intensive to find or create a particular custom component you are after. Also, depending on the complexity (and number) of the objects, they might lead to an increase in rendering time. On the positive side, this method results in consistent renderings if several perspectives of the same scene are created.

- **Add to rendered images using image-editing software (in postproduction)**—You can often easily and quickly create renderings without plants, people, and so forth, and then add them using your favorite image-editing software (e.g., Photoshop). As long as you master selection tools, brushes, and masks, you can very effectively add a large number of entourage elements to a rendering without making the actual rendering process too complex and time intensive.

  A downside of this method is that the same elements might have to be inserted into various renderings, multiplying the workload in postprocessing.

The upcoming sections mention some of the techniques you can employ to include objects in the 3D environment and use them for rendering.

Entourage

When it comes to adding objects to your renderings, your best source may be SketchUp's own 3D Warehouse (see **Figure 5.80**). This virtual warehouse of SketchUp models has grown significantly over the past few years and now boasts an impressive collection of user-submitted objects, brand-name appliances, 3D buildings, trees, and much more. A few years ago, even the makers of SketchUp decided to move the object collection that beforehand shipped with SketchUp into the Warehouse (leaving only a few sample objects behind in SketchUp's installation folder).

You can find the 3D Warehouse at http://sketchup.google.com/3dwarehouse.

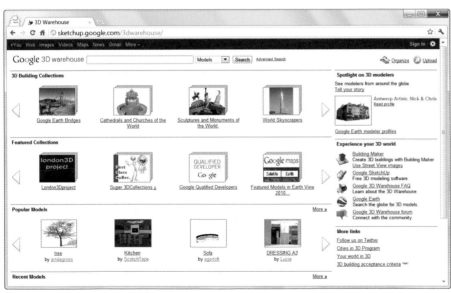

Figure 5.80: 3D Warehouse main screen

You can currently incorporate 3D Warehouse objects into your SketchUp project in a variety of ways:

- **Browse the 3D Warehouse using the "Get Models . . ." button.** This gives you a web browser that opens right within SketchUp and allows you to load any object as a

component into the current model. This method is usually best if a search doesn't immediately return the desired object or if more detail about a model is desired.

- **Search the 3D Warehouse from the Components window.** When you search for any term in the Components window, SketchUp displays the most relevant models it found in the 3D Warehouse. You can then load any component with one click into the current model.

- **Browse the 3D Warehouse using your web browser of choice.** You can alternatively search the 3D Warehouse, download any models you like to your computer (as SketchUp's SKP files), and then import (or drag-and-drop) the components into the current model. This works well if you need to import multiple components.

While the 3D Warehouse may be the largest resource for SketchUp models at this point, it isn't the only one. Various manufacturers offer 3D models of their products for download, and there are also commercial websites that sell well-crafted generic models (e.g., www.formfonts.com). Sample company sources are:

- www.maglin.com/sketchup.html—Landscape furniture.
- http://sketchup.google.com/3dwarehouse/search?uq=02096593275336267876042828&styp=c&scoring=m (Short link: http://goo.gl/6ST1e)—GE has modeled many of its appliances and put them into the 3D Warehouse.

When you work on lots of renderings, it is a good idea to create your own library of these entourage items. That way, you can dress up a standard scene very quickly. Select your favorite SketchUp components for their quality, good rendering appearance, and, possibly, low polygon count, and archive them. Although you could create a library on your hard disk, you can also create a Collection under your account in the 3D Warehouse and add your favorite components to it. Those components will then be available in SketchUp via the "Get Models. . ." button on the Getting Started toolbar.

Urban Context

If your entourage consists of neighboring buildings—as in the case of a model that is to be shown in its urban context—then you can use the same technique as previously described to load those buildings into SketchUp. All you need to do is select the Nearby Models menu option in the Components window, and 3D Warehouse models that are located within a certain radius of your model's location will be offered for download.

TIP	Keep in mind that you can download only user-submitted (or SketchUp-submitted) content from the warehouse. Some buildings that you see as 3D buildings in Google Earth (especially in larger cities) were created by commercial companies for Google and are not available for download. You can identify those buildings in Google Earth when you hover over them and they don't get highlighted.

It is important to note that you must make sure your model is properly geo-located before you can do this (otherwise SketchUp doesn't know where it should search for buildings). To locate your model, select the "Add Location. . ." button from the Getting Started toolbar or the File menu.

Figure 5.81 shows a scene after several models have been loaded into the file.

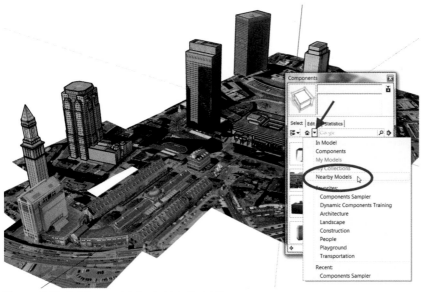

Figure 5.81: Building models loaded into SketchUp

If you prefer to show surrounding buildings without photographic imagery on their facades (i.e., as "clay" or "massing" models), you can accomplish this by either repainting them with a neutral color (you might have to double-click them to be able to edit their "inside" faces) or by deleting all facade texture materials from the Materials dialog. When a material is deleted, it is replaced by SketchUp's default material, and all faces appear white.

People

People are an important component in a rendering. Not only do they add scale to a scene, they also personalize the scene. For example, in a street-view rendering you can create very different neighborhood ambiances by either adding businesspeople in suits or families with strollers. Of course, you want to add people as appropriate, and as necessary for the message you intend to convey.

To add people (or animals, for that matter) to your scenes, you can take any or all of these approaches:

- **Add 3D people to the scene.** These are complete 3D models of people—sometimes even with realistic-looking textures. The 3D Warehouse is a great resource for these kinds of models, but there are also other (sometimes commercial) distributors.

  The main benefit of using 3D models is that once you have "staged" a scene with people, you can create as many realistic-looking renderings from various perspectives as you need. A potential downside of this approach is that some of the models are composed of many polygons and may slow down rendering—especially if you use lots of people in your scene.

- **Use 2D people shapes or cutouts.** SketchUp allows components to be set up so that they always face the camera. This is very convenient (e.g., you need only a 2D photo to represent a tree from various perspectives), but it results in the people in your scene always facing the viewer—no matter the vantage point from which the viewer looks at them.

  You can find many 2D people in the 3D Warehouse. Some are created from photographs and therefore look very realistic. Others are "sketchy" people who look better in sketchy non-photorealistic renderings.

  In terms of rendering speed, these will render faster than true 3D objects. This is true especially if the rendering software allows for transparent image maps (e.g., using a PNG image with transparency) and the object has only a few polygons.

- **Add people as 2D cutouts in your image-editing software.** If you are comfortable using image-editing software such as Photoshop, then you can learn to place people cutouts realistically into any image. While this may require picking appropriate photos and some postprocessing work on shadows and layering (e.g., when a person stands behind a plant), this approach is very efficient. Because people aren't being added to the rendered scene, they don't add any additional time to the rendering itself.

  A downside of this approach is that it is not very usable when various perspectives of the same scene need to be rendered and the people need to appear consistent in the different perspectives.

Figure 5.82 illustrates some of the people models that you can download from the 3D Warehouse (from left to right): sketchy 2D, silhouette 2D, photo-cutout 2D (model by TaffGoch), sketchy (untextured) 3D, and photorealistic 3D (model by Reallusion iClone).

Figure 5.83 shows how the three main types—2D follow-me, textured 3D, Photoshop-added image cutout—behave when two views of the same scene are being rendered. As you can see, the only one that looks appropriate in both views is the 3D model of the person in the middle. The left figure renders facing the camera independent of the viewer's location, which is convenient but may not be desired. The figure on the right works only in the top image because for the bottom image a different photograph (one taken from behind) would have to be available.

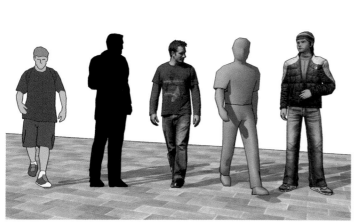

Figure 5.82: People models available from the 3D Warehouse

Figure 5.83: Different perspective renderings of people

Example 5.5: Adding a Person to an Image in Photoshop

To add a person using your photo-editing software, follow the procedure illustrated in **Figure 5.84**. These instructions are for Adobe Photoshop—modify them slightly if you are using another type of software.

1. Load the rendered scene image into Photoshop.
2. Load a photograph of a person into Photoshop. Make sure the view and the lighting roughly work with the scene and viewpoint of your rendering. It is best if the photograph has a uniformly colored background because you need to remove this background in the next step.
3. Cut out the shape using appropriate selection tools (the Lasso and the Magic Wand tools are very helpful here).
4. Paste the shape twice into your rendered image. You create two different layers this way.

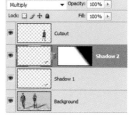

Figure 5.84: Adding a person and her shadow using Photoshop

5. Turn off the topmost layer, which is the image of the person that will be visible in the end.

6. On the second layer, color the shape completely black.

7. Select the shape and stretch it until its perspective makes it appear as the person's shadow.

8. Apply a Gaussian blur filter to this layer. Go with a light blur for this one.

9. Copy this layer and apply Gaussian blur again—this time with a higher degree of blurring.

10. If you want a shadow that appears less fuzzy next to the person yet blurs farther away, apply a layer mask as shown in **Figure 5.84** to the two shadow layers.

11. Make the person's image visible again and place it on top. It should now look correct in the rendering.

Plants, Trees, Grass, Rocks, and Carpet

It is easily possible for computer-generated geometry (be it the model of a house, a landscape, or simply an object) to look artificial. In our SketchUp models, edges are always straight and most lines are parallel to one of the three main axes. As you know, nature doesn't look like that. When you stand in front of your house, you will see not just the straight lines of the facade but also a tree, a bush, some grass, some dirt, and other "natural fuzziness." It is therefore important—even if landscape elements are not your main focus—to include them in order to add realism. This is, of course, even more important when you focus solely on the landscape and need to not only show vegetation but also consider species, growing season, and so on.

While this is very important for exterior scenes, interior scenes often benefit from adding plants as well. From a rendering perspective, adding a shag carpet and some fabrics (drapes or bedsheets) serves a similar purpose. Chapter 4 showed you how to use the Fur plugin to create a carpet. We therefore focus on vegetation in this section.

2D versus 3D versus Photoshop

When we want to add landscape objects such as plants to a scene, we face the same question that we did with people: Should we include complete 3D models in our scene? Can we get away with 2D (face-me) images? Or should we add them in postproduction (using Photoshop or any other image-editing software)?

As before, this often depends on various parameters—in this case, the most important is how close an object is to the viewer. The closer the object is to the camera, the more detail is visible and the more modeling might be necessary to include that detail.

Another aspect to consider in our model is polygon count. Natural objects are always crooked, random, edgy, or at least bent (imagine a blade of grass). Modeling this in detail often requires many polygons to make these objects look realistic. We might be able to use fewer polygons and add detail using a texture, but that is not always possible.

In general, follow these recommendations:

- **Object is close to the camera**—Model as much as necessary. You might be able to model only one side of the object (e.g., the front side of a wall). Use as few polygons as possible and try to add randomness by using a texture. Prefer a single-image texture rather than a tiling image (tiled images often create patterns that can easily be spotted up close).

- **Object is far away**—Use a 2D image of the object if you have one and its alignment works. This is a great method for adding trees as long as you can create some random variation in them (by scaling, etc.). Of course, the image must have transparency so that it doesn't include a background of its own (PNG and GIF file formats work well for this). Alternatively, use a coarse 3D model. You might be able to use simpler textures or just colored materials. If you use texture images, try using tileable textures, such as the ones included with SketchUp.

Figure 5.85 illustrates this approach. While the tree in the front is modeled in 3D at a reasonable level of detail, the tree in the back is a face-me photo-based 2D image. The sky and landscape beyond the hill come from a sky probe—you could even use it to add more trees, or other landscape objects, in the background.

Figure 5.85: 2D and 3D trees in the same scene

Adding 3D Plants

Let's take a look at how to add 3D plants and make them look good in a rendering. We'll add a tree here, but this works analogously for shrubbery, flowers, and the like.

For this scene, we will use one of the trees that come from SketchUp's 3D Warehouse. You can find it as "Aspen deciduous tree high polygon." The top-left image in **Figure 5.86** shows how it looks when you insert it into your model. As you can see in your model, the trunk is textured but the leaves are simply colored. For my purpose, I replaced the trunk texture with a browner one. This is not correct for the species but works better for now.

When you render the tree as is, you should get a result similar to the top-right image in **Figure 5.86**. Although this already looks quite good, it could use some refinement. The leaves are green, but they look very flat. Also, their color is a bit too stark to look natural. Furthermore, although the trunk has a good texture, it also looks flat. To improve the rendering, I made these changes:

- **Leaf texture**—To add a bit more texture to the leaves, I added a tileable leaf image to their material. You can see it in a bit more detail in the bottom-right image.

Figure 5.86: Adjusting a tree for rendering

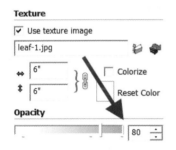

- **Leaf transparency**—When you hold a leaf up to the sun, it lets light through. Therefore, in SketchUp, it also helps to add a slight transparency to the leaves. In the Materials window, adjust the opacity slider to 80%.

- **Leaf reflectivity**—In my rendering software, I gave the leaf a slightly reflective preset. Don't go overboard with this setting—unless you are modeling a rubber tree (or wet leaves), of course.

- **Leaf and trunk bump**—As always, some material bumpiness helps a lot. This is true especially when we have something as coarse as bark. I gave both the leaves and the trunk texture a bump, reserving the higher magnitude for the trunk. You can see the result of this in the bottom-right image.

You can evaluate the result of these setting changes in the bottom two images in **Figure 5.86**. You can also see that, although the tree model works well at the distance shown in the bottom-left image, the bottom-right image is too close and would require some additional modeling of secondary branches. It is also not realistic that leaves intersect the stem—this would need to be fixed, too.

Adding Grass

Grass is a rendering component for which quality and preparation effort are very much dependent on the distance between the viewer and its blades. Imagine looking at a lawn from several houses away. The lawn (at least, if it is well maintained) looks like a uniformly colored and textured green plane. As you get closer, you see the individual blades of grass—especially at the edges of the lawn, where blades overlap walkways. If you bent down to get even closer, you would then see that not only is each blade curved, it also has a V shape in cross section. As you can see from this example, your approach to modeling and rendering must vary, depending on the distance of the viewer from the object.

Figure 5.87 illustrates the four approaches you can take.

Figure 5.87: Grass rendering methods (left to right: SketchUp's grass texture, same with bump, individual blades, 2D image patch)

When your distance to the grass is quite far, it might be enough to simply use SketchUp's grass texture (or any other uniformly patterned tiling grass texture). This is shown in the left-most image of **Figure 5.87**.

This method improves when you add a strong bump to the grass material in your rendering software. The deep shadows that appear give the grass more realism, as you can see in the second image from the left. You can evaluate the appropriateness of this approach when you are at arm's length from the image—this rendering should appear reasonably realistic.

The third image shows the other end of the effort spectrum. This is what we get when we model each blade separately. I used the Fur plugin (from Chapter 4) to do this and went with its default arch shape. As you can see in the third image from the left, this approach leads to an excellent degree of realism—even when the viewer is as close as we are to this image. A major downside of this approach, however, is the very large number of polygons that this method creates, which not only increases rendering time but can also slow SketchUp significantly if you are not careful. Use it mostly for close-ups.

A good compromise between those two approaches is to use a 2D image of a grass patch (see **Figure 5.88**), which contains a transparent background (a transparent PNG, in my case). Especially when you use a low-resolution image, this approach can significantly reduce the polygon count in your model. I also used the Fur plugin, as before, but this time let it place copies of the image component instead of individual blades. You can evaluate the result in the far-right image in **Figure 5.87**.

As you can see, this leads to quite a good rendering. Each grass-patch image (representing many blades) triangulates as two polygons, while each individual blade in the last method turned into five—making this method much more polygon-efficient. The only problem with this approach are occasional patches (you can see some in the foreground) and a bit of a "scruffy" appearance. Because these images are all vertically placed, it is also not possible to get a good result when you look down onto the grass. Nevertheless, this approach is very efficient, and any problems can easily be fixed later in your image-editing software.

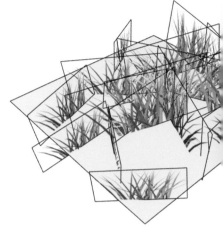

Figure 5.88: Using transparent images for grass

Figure 5.89 combines both approaches. The blades close to the camera were modeled as single blades (with a striped green texture), while the grass in the background consisted of randomly placed images.

Figure 5.89: A grassy hill

Example 5.6: Making a 2D PNG Cutout Face-Me Component

Follow this procedure to make your own 2D cutout SketchUp components for trees, people, and so forth. This example uses a tree image from www.cgtextures.com and turns it into a face-me component.

To be able to remove the background in an image, it is best to start with a uniformly colored background. If you are taking the pictures yourself, look for a white wall, a gray sky, or any other uniform texture as a backdrop. To make sure the image can be used as a component, take it straight on, not looking down on or up at the main subject. It is also a good idea to take the image when the sky is overcast—this eliminates any unsightly shadows in the component.

1. Let's start with the raw image. Note that it has a relatively uniform background.

2. Load it into your favorite photo-editing software. This example demonstrates the process in Photoshop, but you can do this with other software, too (e.g., Gimp).

3. Remove the sky background and the portion that includes the hills. Use the Magic Eraser tool, set tolerance to 50, check Anti-alias (to smooth edges), uncheck Contiguous (so that areas inside the tree will also be removed), and keep Opacity at 100%. Next, click anywhere in the sky area. This removes the sky but not the hills in the background. To remove the hills, click anywhere on them. The result should look like the image shown here.

As you can see, the background has largely been removed and replaced by Photoshop's default checkered pattern.

4. Let's check how well we did. Add another layer in Photoshop, fill it with a dark solid color, and place it underneath the tree layer.

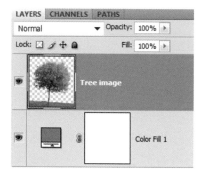

5. As you can see in the image, the background was removed quite well in the darker area (at the bottom left), while the right side of the tree still has a slight white "halo."

6. On the tree layer, use the Magic Eraser tool again. Experiment with the tolerance setting and click at the right edge of the tree until just the white halo is removed.

7. Now we will focus on the base of the tree. To remove the grass and other ground, use any combination of the Eraser, Magnetic Lasso, and Magic Wand tools that works well for you. Because the tree trunk is reasonably straight in this image, I was able to just erase everything around it.

8. Use the Eraser tool to cut the base of the trunk parallel with the ground and clean up the image as necessary. This is what you should have now.

9. Crop the image to the edges of the tree, scale it to your liking, and save it as a transparent PNG file.

10. In SketchUp, change your view to a parallel-perspective front view, which reduces errors in the next steps.

11. Import the image into SketchUp using the **File → Import. . .** menu item. Make sure you check "Use as image" in the import options.

12. Double-click on the origin to place the tree in your model. Scale the tree to your liking.

13. Right-click on the tree image and select Explode from the context menu. This turns the image into a rectangle with the tree image as its texture.

If you want the tree to have appropriate shadows in SketchUp, trace its outline with the Pen tool and cut away the remaining parts of the rectangle using the Eraser.

14. Before we turn the tree into a component, we need to hide all the bordering edges. Do this by holding Shift while clicking the edges with the Erase tool. This hides them instead of erasing them.

15. Select the tree again and click on the Make Component button in the toolbar (or use the right-click menu). In the dialog that comes up, make sure you check the "Always face camera" box. Don't click on the Create button yet!

16. In the same dialog, click on the Set Component Axes button to place the axis at the base of the tree. This will be the insertion point for the tree component.

17. When you click on Create, your new face-me tree component has been created (see **Figure 5.90**). Orbit around your model and see that the tree component behaves properly. Also, add a few more component instances from the Components window (or by copying them).

Figure 5.90: Finished face-me tree component in SketchUp

Rendering Tips

- **Keep your scene simple, well modeled, and organized!** These are three important principles to reduce the chance that rendering your SketchUp scene will give you a headache. Here's what I mean:

  - **Keep your scene simple.** Model only what you absolutely need for your rendering. Often, there is no need to model geometry that is out of view or inside an object. If your model is complex, it might be a good idea to make a copy of the file and then prepare the copy for rendering (by deleting and cleaning items) rather than using the original.

    Delete any face that does not contribute to the rendering. Make sure there is no leftover geometry (e.g., from a previous intersection operations).

  - **Craft your model well.** You will run into trouble if you have small holes between surfaces, overlapping surfaces, or "crooked" planes.

    Any gaps in your model increase the chance of light leaks—bright spots where there should be none, as, for example, in a corner. If your scene has overlapping surfaces, intersect them and trim excess geometry. And if you have planes that should be flat but aren't because one vertex is out of plane, this surface will be triangulated during rendering, and you might see edges appear in your images.

  - **Keep your model organized.** If you use many copies of the same object, make sure they are components rather than groups. Many rendering software packages nowadays are able to handle object instancing just as SketchUp does. This means that instead of exporting lots of faces per copy, these tools use the more efficient method of component definitions and instance locations.

- **Reduce the number of polygons.** Every additional polygon increases rendering time and makes your model more complex. This is true especially if you assign translucent, reflective or light-emitting materials to the polygon. Therefore, keep your model's polygons to the necessary minimum.

  This is the case particularly with natural objects such as trees or grass. When they are at a certain distance from the viewer, it is often a good idea to either reduce their complexity or replace them with a photograph of the object.

  Also keep in mind that while SketchUp is able to handle multivertex polygons, rendering software usually uses only triangular polygons and triangulates any nontriangular face coming from SketchUp.

- **Use fully reflective materials or fake reflections.** Reflective surfaces in your scene increase rendering time, especially if they are just partially reflective (such as brushed aluminum). Unless you absolutely need this effect, try going with full reflectivity (a mirror or highly polished chrome material) or a gray gradient texture (faking blurry reflectivity) instead. This works especially well when the object is not too close to the viewer.

- **Consider using fake caustics.** If you need caustics (the light ripples that reflect off a pool, for example) but don't want to expend the additional rendering time that caustics require, consider faking them by projecting a caustics texture image onto a wall, ceiling, or other surface.

- **Handling depth of field, light rays, and motion blur.** All of these effects increase rendering complexity. Consider adding them in postproduction using your photo-editing software's filters.

  Depth of field can usually be added using Gaussian blur, and motion blur is a filter that is typically available in your editing software. Experiment with the settings and master masking techniques to perfect these effects.

  Light rays (as you would see coming through trees on a slightly foggy day) require your rendering software to simulate particles in the air (this is often called *volume* lighting). They can often be added in postproduction, especially if they are produced by simple geometry (a window, for example).

- **Use postproduction to your advantage.** Depending on your image-editing skills, it might be more efficient to add people, vegetation, sky effects, and some shadows in your favorite photo-editing software after the rendering is done.

  A great resource for learning and discussing techniques is this website: www.sketchupartists.org.

Making Renderings Presentable

While photorealistically rendered images can be produced with a quality similar to that of actual photographs, sometimes this is not desirable. For example, you might want to create sketches that have a watercolor appearance but contain effects like lighting or reflections that you can produce only using rendering software.

To this end, it is often useful to expend as much effort on your renderings in postproduction as you do in setting them up. The following examples give you some ideas for paths to follow.

Combining SketchUp Output and Rendered Output in Photoshop

Since SketchUp has very good built-in support for hand-drawn linework through its styles, adding a "handmade touch" to your renderings can easily be accomplished by exporting a sketchy image of your current model in addition to creating a rendered image.

Let's start this process by using the rendering shown in **Figure 5.91** as it came from the rendering software.

Figure 5.91: Raw rendered image

Although the grass and the soft shadows look good already, you might still want to add some sketchy linework to the image. To do this, try out various styles in SketchUp until you find one you like. Let's use the one in **Figure 5.92**. Export it using the **File → Export → 2D Graphic** menu item.

Figure 5.92: Sketchy view exported from SketchUp

Alternatively, you could also have printed the SketchUp view, overlaid it with tracing paper, and sketched onto it yourself. Then just scan the sketch, scale it and overlay it onto the rendered image in Photoshop if you go this route.

TIP	Create both images—the rendering and the SketchUp exported sketch—at the same resolution. This makes it easier to combine them in your image-editing software.

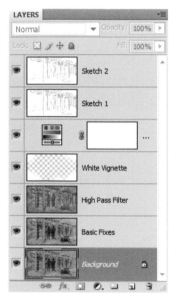

Figure 5.93: Layer arrangement in Photoshop

To combine the two images, I imported both of them into Photoshop and used the following layer arrangement (see **Figure 5.93**). As always, if you use different image-editing software, adjust this appropriately.

Let's go through the layers, from bottom to top:

- The Background is simply the rendered image.
- The Basic Fixes layer is a copy of the background in which I cleaned up some things (moved some knots around for diversity) and applied the Diffuse Glow filter.
- The High Pass Filter layer is a copy of that layer. I desaturated it (to make it gray-scale) and applied the high-pass filter. I then applied the Soft Light blending mode to this layer, which basically uses this layer to make contrast a bit more interesting.
- The White Vignette is a transparent layer onto which I painted white edges and which has an opacity of 70%. This fades the image out nicely toward the edges.
- The next layer is a Hue/Saturation adjustment. I reduced saturation a bit because colors were a little too bright.
- Next are two copies of the sketch. The lower layer of the two has a Soft Light blending mode (to soften colors even further), and the top one has a Multiply blending mode (to add crisp linework).

The image created from this process looks like **Figure 5.94**.

Figure 5.94: Combination of sketchy and rendered image

Another example for this process is the cover image of this book where a layer-blending technique was used to transition between a sketchy "SketchUp" image of the building (shown on the left) to a photo-realistically rendered one (shown on the right).

Other Methods

The method just described is a good staple to have in your tool kit. In addition, you can try the approaches shown in **Figures 5.95**, **5.96**, and **5.97** (and combine any of them as necessary). Some of them use the free FotoSketcher software that you can download from www.fotosketcher.com. This software lets you convert an image into a sketch, a watercolor or an oil painting (complete with brushstrokes) that you can then layer onto a rendering.

Figure 5.95: Fading out background using blurred layer and layer mask based on fog image from SketchUp

Figure 5.96: Watercolor using FotoSketcher

Figure 5.97: Painting using FotoSketcher

Your Turn!

The following tasks allow you to practice the topics in this chapter:

1. **Model a photo**
   Take a picture of a small setup (one that is easy to model in SketchUp). For example, lay a metal knife on a wood block on a granite counter and take a close-up picture that shows all three materials. Evaluate the materials in the photo with respect to reflectiveness, bumpiness, irregularities, and other qualities. Then model the scene in SketchUp and re-create it in your rendering software. Your main challenge will be to re-create material properties and the lighting environment as closely as possible to the real world. Render the scene and compare it side by side with the photo.

2. **Create a loft sales brochure**
   Model one or two typical loft scenes (a living room with a sofa and artwork on the wall or an open-floor-plan kitchen). Then apply good-quality materials, set up the scene with interesting lighting, and render it. Finally, put the images together on a single page, add some text, and turn it into a sales flier for a new downtown loft.

3. **Day and night**
   Using the scene from the previous task, create good-looking night renders (complete with artificial lighting). Make sure the night renders don't just have a black sky. Then present the day and night renders side by side.

4. **Visual styles**
   Using any method described in this chapter, create three visual styles that cannot be done only by using renderings, SketchUp's Styles, or FotoSketcher. Use any rendering you like as basis for this.

5. **Bring yourself into SketchUp**
   Get someone to take a photo of yourself (or a friend) and turn it into a cutout face-me component for SketchUp.

Chapter 6
Creating Geometry
Using Ruby Scripting

In this chapter, we will create scripted geometry in SketchUp. Often called *computational geometry,* this is an exciting current topic that allows you to produce 3D shapes and arrangements that are impossible (or at least hard) to create by hand. Scripting in SketchUp also allows you to automatize repetitive tasks or facilitate data exchange with text files.

Key Topics:

- What is computational geometry?
- Using Ruby Scripts in SketchUp
- Working with multiline editors
- Introduction to Ruby
- Overview of SketchUp's API
- Creating scripted geometry
- Transforming geometry
- Modifying geometry based on attractors

Why Computational Geometry?

Even to the most seasoned SketchUp user, some tasks are challenging to complete by hand, using only the tools SketchUp provides. A good example is the creation of copies of a component where all the copies need to be placed at locations that vary not only linearly (e.g., "place a copy every 16 inches") but also in intervals that are determined by a formula (e.g., where the y coordinate is determined by the square of x—which, of course, results in the shape of a parabola; see **Figure 6.1**).

Another example is a curved building whose window-shading devices need to be designed such that the degree to which they shade or open is determined by how far they deviate from the southern direction. (See **Figure 6.2**.)

Such tasks can sometimes be accomplished using one of the available plugins. However, those plugins are usually limited to the supplied feature set and are hard or impossible to modify to your needs. As it turns out, some complex tasks may actually be easier to script in a few lines of code than to model by hand.

Figure 6.1: A parabola-shaped park

Figure 6.2: Parametric solar shading

Fortunately, for the user, both versions of SketchUp (the free and the Pro version) come with a scripting language preinstalled and readily available. Also, when it comes to scripting, there is almost no difference between the feature sets of the free and the Pro version, providing the user of either version with a significant tool set.

The scripting language SketchUp uses is called Ruby. This language, created in the mid-1990s by Yukihiro "Matz" Matsumoto in Japan, might be unknown to many software users (for example, Microsoft Office products usually ship with another language called VisualBasic and websites typically use a language called JavaScript). However, it has been a very popular language among Web application developers, and many large and popular websites (Twitter,

Groupon, and Hulu are among them) run on code written in Ruby. Besides, it is a very object-oriented language that is actually quite easy to learn, as you will see. (See **Figure 6.3**.)

SketchUp itself provides a large variety of tools and procedures that are available to the Ruby scripting language using what is called an application programming interface (API). This gives you access to tools such as Move or Copy, as well as the entire collection of materials in your SketchUp model, through many well-documented functions.

Having said all of this, it is important to note that any tool or method has its limitations. Consequently, scripting in SketchUp also has limitations. For example, curved shapes are not implemented as perfect curves, in SketchUp's tool set nor in the API. Because SketchUp is a polygon-based modeling software, it

Figure 6.3: The Ruby language's website

naturally does not have a way to define perfect curves. As you know, circles are always made up of several polylines in a SketchUp model. While this certainly is a limitation, it is also important to keep in mind that in real construction, curved shapes may be approximated by planar elements (e.g., plane glass panes on a curved-looking facade)—often for constructability and cost reasons. As a result, SketchUp's API does not natively support NURBS (nonuniform rational B-splines)—at least not without a plugin. This is better implemented in other software programs (McNeel's Rhinoceros and Autodesk's AutoCAD, Revit, and Maya, for example).

Nevertheless, creating scripted geometry is quite possible using the freely available version of SketchUp, which lowers barriers of entry and offers the potential for everyone to dabble in this exciting field and create advanced geometry.

This chapter approaches SketchUp scripting in a pragmatic way. This means that every step in the process of geometry creation and manipulation does not necessarily have to be scripted—after all, we are not writing a plugin, and SketchUp already has a great set of built-in modeling tools. We'll script only what can't efficiently be modeled by hand, using standard SketchUp modeling tools to start or finish a project.

First, let's set up your computer so that you can easily script using Ruby.

Setting Up Your Computer

Before we do any modifications to SketchUp, let's use Ruby as SketchUp provides it to get you started quickly. In SketchUp, look for a menu item in the **Window** menu called **Ruby Console**. When you click on it, you should see the dialog shown here.

This is the built-in interface for the Ruby scripting language. It works simply by accepting a line of code in the text box at the bottom. Once you hit Enter, the line is executed, and the result appears in the field at the top of the box.

Let's try this out: Enter `puts "Hello World!"` into the text field and hit the Return key. The result shown here should appear in the dialog.

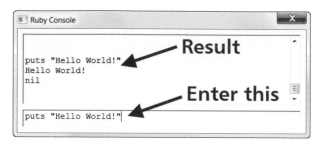

Because the `puts` command simply prints something to the result part of the dialog, all you get as a result is the "`Hello World!`" string.

You can even get a little more "programmy" by assigning variables. First, enter `a = "Hello World!"`, hit the Return key, and then enter `puts a`. You will find that you get the same result. In this case, however, we assigned a text string to a variable first.

Let's try something a bit more useful—let's do some calculations. As it turns out, you can essentially use this dialog as a built-in calculator. Enter `2*16`, and you should get a result of 32. You can use any of the common mathematical symbols here such as `+`, `−`, `/`, `*`, and `**` for power (as in `2**4` equals 16) as well as parentheses to structure calculations.

To make things more interesting, let's try a square root. Enter `sqrt(10)` and you should see `3.16227766016838` as the result. Nice, isn't it?

As you might have guessed, you can use any of the usual mathematical operators and functions here. This is the complete list of all built-in math functions: `acos`, `erfc`, `asin`, `atan`, `cosh`, `sinh`, `tanh`, `acosh`, `asinh`, `atanh`, `exp`, `log`, `log10`, `sqrt`, `atan2`, `frexp`, `cos`, `ldexp`, `sin`, `hypot`, `tan`, `erf`.

TIP

When using any trigonometric functions such as `sin` for the Sine of an angle, keep in mind that the argument must be in radians. You must add `.degrees` (including the dot) to the angle to get the correct answer. Example: `sin(45.degrees)` results in 0.707.

Furthermore, although these functions should work without a problem, their complete "name" is, for example, `Math.sin`, where the class name "Math" (in which these functions are defined) is prepended (with a dot as shown). Use this syntax instead if you run into any problems.

While this is all great, the preceding examples use only the scripting language Ruby itself. Let's try an example where we actually get SketchUp to do something. Enter the following line of code into the Ruby Console:

```
Sketchup.active_model.entities.add_face [0,0,0],[0,10,0],[0,10,10]
```

Once you hit Return, you should see a triangle appear in SketchUp's modeling area (at the origin, to be exact). It should look like **Figure 6.4**.

You can even access some of SketchUp's user interface components. Just for fun, let's open a file selection dialog. Enter this code:

```
UI.openpanel 'Select any file please'
```

As you can see, this one-liner opens a dialog, lets you select any file, and returns its location on your computer (the complete path) in the results part of the Ruby Console. Don't worry—nothing will happen to the file. All you did was ask yourself for a filename.

As you can see in these examples, the Ruby Console is already very powerful. One thing that it can't do, though, is execute multiline scripts. For this, we need to enhance SketchUp with a plugin.

Installing the Ruby Code Editor Plugin

Throughout the rest of this chapter, we use the free Ruby Code Editor plugin to edit and run Ruby code within SketchUp.

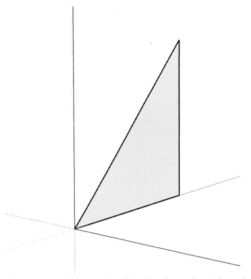

Figure 6.4: Adding a triangle in SketchUp using Ruby code

While this plugin is a convenient way to accomplish the task, there are also other methods and plugins that work well—they will be described later.

Here's what you need to do:

1. Go to my website and download the installer file for the plugin. Download the ZIP or RBZ-version of the installer file (see Chapter 4 regarding how to install RBZ files). www.alexschreyer.net/projects/sketchup-ruby-code-editor.

2. If you downloaded the ZIP-file, open the file using your favorite ZIP compression software. Often, your operating system simply opens the file if you double-click it. If that doesn't work, use software such as 7-Zip (download from www.7-zip.org) to unzip it.

3. Paste the contents of the file into SketchUp's Plugins folder on your hard disk. You can find that folder in the following locations. **Important:** Retain the file and folder organization as it was included in your ZIP file. *Windows:* C:\Program Files\Google\Google SketchUp <version#>\Plugins\ *Mac:* /Library/Application Support/Google SketchUp <version#>/ SketchUp/Plugins/.

4. Restart SketchUp. You should see the following menu item in the **Window** menu (next to the Ruby Console menu item you used earlier). Click on it to start up the code editor.

Find and install Plugins...
Ruby Code Editor
Ruby Console Pro

The Ruby Code Editor starts up as a dialog in SketchUp. **Figure 6.5** shows what it should look like on your screen.

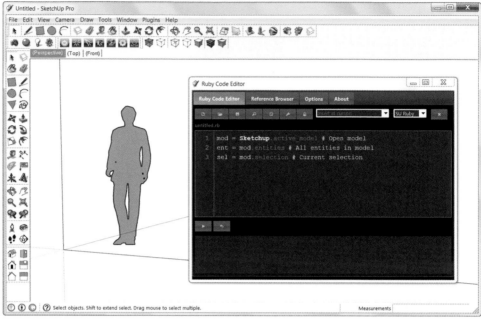

Figure 6.5: Ruby Code Editor

You can try it out by entering any of the code examples that we worked with before into the editor area. In the example shown in **Figure 6.6**, we insert the line of code that adds a triangle to the model (1), press the Run button (2), and the triangle appears on the screen (3). If you don't like what you just did, then simply click the Undo button next to the Run button to undo the drawing of the triangle.

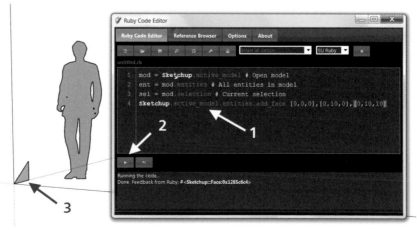

Figure 6.6: First example in the Ruby Code Editor

As you can see, the Ruby Code Editor has a few advantages over SketchUp's Ruby Console. For example, you can enter multiline text, source code is highlighted appropriately, and you have Run and Undo buttons. Furthermore, you can save and open code snippet files and even work on files in HTML and other languages. There is also a handy drop-down that lets you insert common code snippets at the current cursor position. For now, we'll ignore some of the more advanced features, however. Just peruse the editor and discover some of its tools on your own terms.

Now you are ready to tackle all of the examples in this chapter.

Other Options: More Plugins and External Editors

There are some alternatives (all free) to my Ruby Code Editor plugin that have more or less the same functionality. All of them work fine with the examples in this book.

- **WebConsole**—This plugin written by Jim Folz was the original multiline editor on which I based my editor. You can download it from here: http://sketchuptips.blogspot.com/2007/08/plugin-webconsolerb.html.

- **Ruby Console Pro**—This plugin was written by Martin Rinehart and offers a multiline editor as well as a script environment, which provides shortcut functions for common tasks. http://forums.sketchucation.com/viewtopic.php?t=28571.

If you prefer not to use plugins at all, or if you like using an external code editor, then you can use this method:

1. Using any text (or Ruby code) editor you like, create a text file `source.rb` (or similar) in any convenient location, such as in the root of the C-drive.

2. Add some Ruby code to it and save the file.

3. Open the Ruby Console in SketchUp and type the following line when you are ready to run the code: `load 'C:\source.rb'`.

4. This loads all code from the file and executes (runs) it line by line.

TIP

If you prefer the external-editor method and you become more familiar with writing Ruby code, then it might be a good idea to add the load script to SketchUp as a menu item (using a small plugin that you will have to write). This gives you one-click access to the function.

Another (more advanced) method is to use the free **SketchUp Bridge** tool that you can download here:

http://plugins.ro/labs/bridge_tut.htm

Intro to Ruby and the SketchUp API

This section gives you a primer for the Ruby language as well as an overview of the most important aspects of SketchUp's API. If you need more in-depth information, follow any of the links at the end of the section.

For the impatient: If you prefer to just get things going without learning the basics of the language first, then skip this section and go to some of the code samples in the next section (starting on page 245). Any of them work as they are when you run them in SketchUp. You can always come back later and read up on the basics here.

How Does Ruby Work?

Objects, Objects, Objects

Ruby is essentially an object-oriented language. Everything in the language is an object, and every object has properties and methods. You can always apply any appropriate method to an object. You can even add more methods, if you like. Here's what I mean.

Assume you have an object `an_object` and the object has a method `a_method` that does something with it. You can then apply the method to the object simply by adding it with a dot as in `an_object.a_method`. If you want to apply more than one method, you can simply chain them, as in `an_object.a_method.another_method` (as long as they are compatible). Here are some examples that make this a bit clearer.

"`Hello World!`" is a string, as we found out earlier. It is also a String object that has some built-in methods. One of those is `length`. You can therefore enter "`Hello World!`".`length`, which yields 12, the length of the string in characters.

An example for chaining methods is `125.to_s.length`, which gives the result 3. Now, what happened here? We started with `125`, which Ruby automatically interprets as a number (an integer, actually). We then applied a conversion method to it (`to_s`), which converts the number object to a string. Finally, we added the `length` method that tells us the length of the string in characters, which, of course, is 3.

A similar operation in SketchUp is converting from one unit system to another. As you know by now, internally SketchUp uses inches as its base unit, no matter which localized version you are using. If you want to convert a number in inches to, say, centimeters, you can simply use the built-in numeric function `to_cm` to get the correct value. Therefore, entering `1.to_cm` results in `2.54`.

As you learn more about methods, you will likely come across these two strange-looking types of methods (shown here as examples):

- `String.upcase!`—A method with an exclamation mark at the end modifies the original object in place (in this example, we turn a string into uppercase). Therefore, you don't need to assign it to another variable.

- `String.include?`—A method with a question mark checks something on the object and results in a `true` or `false` (Boolean) answer. The example method shown here checks whether a string includes a substring.

One important principle to understand about object-oriented programming languages like Ruby is that objects (and the classes that define them) have one or more ancestors (*parents*). Analogous to the way we receive our parents' genes, objects in Ruby receive (*inherit*) their parent's methods and properties. For example, the `Edge` object in SketchUp, whose parent is the `Drawingelement` object, has a `bounds` property that it inherits from that object, although by itself it doesn't have this property (as you can see in Appendix C). This may sound confusing at first, but you will soon find that it is a very useful principle.

Storing Stuff: Variables and Constants

As demonstrated, you can simply assign anything to a variable name of your choosing (e.g., a = 1.2). This works with a string, a float number, an integer, an array, or an object. You don't have to even declare the variable's type as you often need to do in other programming languages.

TIP

When using variables, start the name with a lowercase letter or an underscore ("_"). Uppercase first letters are reserved for methods and constants (e.g., for PI). You can assign constants, then, such as MYNAME = "Alex".

You can even mix types, for example, in your calculations:

```
a = 1.2      # a float number
b = 2        # an integer number
a * b
```

This correctly gives you 2.4 as the result. Just keep in mind that numbers stored as strings don't automatically convert in your calculations; use to_f (for float numbers) and to_i (for integers) first. If you had started the three lines of code with a = "1.2" (with the quotation marks, thereby defining this as a string), the calculation would not have worked. The last line would then have to be changed to a.to_f * b instead.

TIP

A useful constant is __FILE__ (with two underscores on either side of the word FILE). It always gives you the current Ruby file's name and location.

It is important to remember that these variables are local variables. This means that they work only wherever they were declared (within a program or a function). If you need a variable to work globally, add a dollar sign to the front: $a.

If you want to use a collection of items, it is often best to store them in an array. An array can, for example, contain a set of coordinates, names, or values. There are two ways to add elements to an array. Either do it when you create the array:

```
names = ["Michael","John","Agnes"]
```

or first create an empty array and then add elements (to the end):

```
names = []
names.push ("Michael","John","Agnes")
```

If you want to access an element of an array, write (in our example) names[1] to get "John". Always remember that an array starts at index zero!

As in other programming languages, Ruby has a large set of useful functions to accomplish a whole lot of operations on arrays, such as sorting and reversing (see Appendix B for a handy reference).

In addition to "classic" arrays, Ruby can store key-value pairs of information in a special kind of array, a *hash*.

```
grades = Hash["Mike" => 88, "Lucie" => 99]
```

Some String Peculiarities

If you use single quotes (as in 'Hello') to define a string, the text will be output as is. If you use double quotes (as in "Hello") to define a string, any inline variable (as in #{myvar}) or command (such as the newline command \n) within the string will be interpreted. Also, if your text contains quotes themselves, you must escape them with a leading backward-slash \. Here is an example that illustrates this:

```
author = 'someone'
puts 'This is #{author}\'s text.\nWho wrote it?'
```

will output This is #{author}'s text.\nWho wrote it?. If you use double quotes instead:

```
author = 'someone'
puts "This is #{author}\'s text.\nWho wrote it?"
```

this will be the result:

```
This is someone's text.
Who wrote it?
```

As you can see, the newline was interpreted as such, and the author variable was inserted into the string. Other useful escape strings are \t for inserting a tab and \\ for getting a single backslash.

If you want to use a formatted string, use the format or sprintf function, as in:

```
sprintf("%s: %d", "Value", 123)
```

This outputs Value: 123. As shown here, you can easily assign variables to placeholders this way. This is very useful if you want to write a formatted string to a file.

Reusing Stuff: Declaring and Using Methods

As always, the more you do with what you already have, the better. Therefore, if you reuse the same functionality more than once in your script, it makes sense to "package" it into a method. Ruby does this by simply using the def keyword:

```
def my_function( salutation, name )
```

```
 puts "Hello, #{salutation} #{name}"
 return name.length
end

len = my_function( "Mr.", "Bond" )
puts "Your name is #{len} characters long"
```

As you can see, you can assign attributes (such as *name*) to a method, let it do things, and even return values.

With Ruby being an object-oriented language, the most elegant (and language-appropriate) way to reuse code is to write a class that defines methods and variables and then use it by declaring a new object instance based on this class. Because the main focus of this chapter is to create geometry and to get you started in Ruby, the discussion of classes is intentionally omitted here.

Controlling Things: IF/THEN/ELSE

You may want to do something in your code after checking something else. You can do this easily with the following lines:

```
length = 10
if length < 10
  puts "Short"
elsif length > 10
  puts "Long"
else
  puts "Exactly 10"
end
```

Modify this snippet to your needs. For example, you don't need the `elsif` or `else` parts if all you are checking for is `length < 10`. Also, the complete list of comparative operators is `<`, `>`, `==`, `<=`, `>=`, and `!=` for "not equal." You can combine comparisons using `&&` for "and" and `||` for "or" as in `length == 10 && color == 'blue'`.

There are other ways to use the `if` statement, too. Here are some examples:

```
# used as a statement modifier
puts 'Exactly 10' if length == 10

# used as a one-liner
if length == 10: puts 'Exactly 10' end

# the 'ternary operator' version for assignment
label = length == 10 ? 'Exactly 10': 'Not 10'
```

TIP

You can always use the exclamation mark to negate a comparison, as in `!(length == 10)`.

Beyond IF, other control structures are UNLESS, which works similar to IF and uses the same structure, and CASE, which applies code based on variable switches.

Repeating Things: FOR, WHILE, and More

Whenever you need to repeat code in a loop, Ruby actually has various ways to let you do this. Let's look at some of them. First is the "classic" for loop:

```
for i in 1..10
  print i
end
```

As you can see, this uses a range (1..10) to describe how often this code should repeat. Another version is the WHILE statement:

```
num = 1
while num < 10 do
  print num
  num += 1
end
```

You can also use the elegant each method, as in:

```
(0..10).each { |i|
 print i
}
```

In this form, i is assigned to each element. Although in our case this simply becomes a numeric index, we will use this extensively in SketchUp coding later to iterate through objects.

Finally, the integer object also has a times method that you can employ like this. Any (multiline) code in the curly brackets will be executed 10 times in this case.

```
10.times { |i|
 print i
}
```

Making It Work: Some More Ruby Syntax Details

You've already seen that Ruby does not require you to tell it when a line ends (many other programming languages need a semicolon there). There are more language-specific things you need to know:

- A one-line comment starts with a #. Anything after the #, until the end of the line, is ignored by Ruby.

- A multiline comment starts with `=begin` and ends with `=end`.

- Parentheses are optional in method calls (to enclose the arguments), though it might be a good idea to add them for readability.

- You can often make things easier in repetitive Ruby code by using ranges. A range comes in a form like this: `1..5`, which includes all integer numbers from 1 to 5. A good example is `(1..5).to_a`, which creates an array like this: `[1, 2, 3, 4, 5]`. Other valid ranges are: `0.1..0.9` and `'a'..'m'`.

- There are two ways to create code blocks: One uses curly brackets, `{` at the beginning and `}` at the end; the other uses `do` at the beginning and `end` at the end. In many cases, you can use either one to enclose code (e.g., for an `each` statement).

What Else Is There?

There is more—much more. Particularly when you need to define your own classes or use some specialized Ruby classes and extensions, there is more to discover. But—as I've said before—this is a pragmatic introduction and we'll limit language functions to only what is absolutely necessary to complete our tasks. There are several useful links at the end of this section that will help you delve deeper into Ruby. We will also cover file operations in some of the upcoming examples.

What's in SketchUp's Ruby API?

SketchUp's Ruby API is an extension to plain Ruby (covered previously) and includes objects and methods that are useful in SketchUp. Some of these allow you to interact with the SketchUp application or the currently open model. Others allow you to add or modify geometry (*entities*). There are also methods to attach *observers* to anything in the software. This is useful if you need your code to spring into action when the user changes something. Look through Appendix C to get an understanding of how extensive this API actually is.

The full API documentation can be viewed at https://developers.google.com/sketchup, where you can also find numerous code examples and revision notes. The links in the next section give you even more resources for finding documentation, code samples, and discussion forums.

You can use any of the SketchUp API's objects and methods in your code whenever you like. As long as your Ruby code runs inside SketchUp, there is no need to "include" or define any external files. As demonstrated when you added a triangle to the model, one line is enough to get things started. Let's quickly look at this code again and take it apart:

```
Sketchup.active_model.entities.add_face [0,0,0],[0,10,0],[0,10,10]
```

Here we are accessing the currently open model (the `active_model`) through the `Sketchup` object. If we want to add any geometry in there, we can't simply add it to the model; we need to add it to the `Entities` collection in the model. This makes sense, of course, because everything in a model is somehow organized into collections (there are also `layer`, `material`, styles, tools, etc., collections). Once we get hold of the entities collection, we can add a face using the add_face method.

The add_face method, in turn, accepts 3D points as arguments; because this face is a triangle, we need to supply three points. These points, in turn, are sufficiently defined by an x,y,z coordinate, which we are supplying as a three-member array ([0,0,0] is the point definition for the origin, of course). This is a triplet coordinate array, which corresponds to the location based on the x (red axis), y (green axis), and z (blue axis) coordinates, respectively.

TIP

When you look deeper into Ruby, you will learn that what I call an *object* is often called a *class*. I do this to (hopefully) reduce confusion. Technically, methods and the like are defined in a class in Ruby, which then is instantiated as an object when you use it.

When you start up the Ruby Code Editor in SketchUp (installed using the plugin, as described earlier in this chapter), you see that a few shortcuts are already preprogrammed:

```
mod = Sketchup.active_model # Open model
ent = mod.entities # All entities in model
sel = mod.selection # Current selection
```

This makes it easy to work with some of the basic collections. With these three lines at the beginning of our code, we can now simply add a fourth line to accomplish what we did before:

```
ent.add_face [0,0,0],[0,10,0],[0,10,10]
```

If you have selected any objects in your current model, you could simply access the first selected object using sel[0], for example, to move or copy it.

Instead of going through the list of available objects and methods, the next sections show you an overview diagram and explore some of the functionality using code examples. To serve you as a reference, the complete API object and methods listing is included in Appendix C.

Object Reference Diagram

The makers of SketchUp have produced a useful diagram, shown in **Figure 6.7**, that explains visually the object (or class) hierarchy. Please note that the following are not included in this diagram:

Core Ruby Classes: Array, Length, Numeric, String; **UI Classes:** Command, InputPoint, PickHelper, Menu, Toolbar, WebDialog; **Geom Classes:** BoundingBox, LatLong, Point3d, PolygonMesh, Transformation, Vector3d, Vertex; **App Level Classes:** Animation, Color, Extension, Importer, Options Manager, Options Provider, Set, TextureWriter, UVHelper; **Observers:** OptionsProviderObserver.

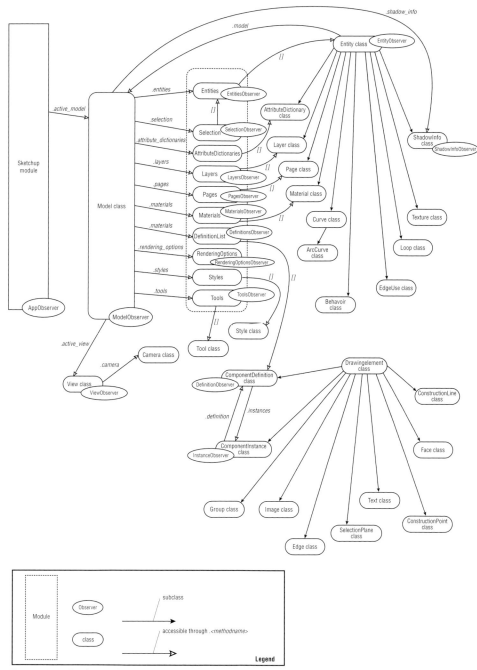

Figure 6.7: SketchUp Ruby API Objects *(Used by permission of Google Inc.)*

Length Units in SketchUp

Because SketchUp works with actual dimensions for lengths, locations, and so on, it is important to understand how to use them in Ruby Scripts, too.

Many examples in this book use plain numbers such as [10,10,10]. Since SketchUp's default unit is the inch (independent of which local version you are using!), these numbers default to inches. Therefore, "10" automatically means "10 inches" in the software.

If you are using a combination of feet and inches or are working with a different unit system altogether (e.g., using meters in the metric or SI system), you must enter units differently. One method is to give SketchUp the length, including its unit, as a string. Fortunately, there is a built-in method (of the Length object) to_l, which attempts to convert any length to SketchUp's internal default unit. Here are some examples:

```
new_length = "1m".to_l
=> 39.3700787401575

new_length = "2'6".to_l
=> 30.0
```

Alternatively, you can use some of the methods of the Numeric object. The following is a complete list of all of its methods:

```
Numeric (Parent: Object)

.cm .degrees .feet .inch .km
.m .mile .mm .radians .to_cm .to_feet
.to_inch .to_km .to_l .to_m .to_mile
.to_mm .to_yard .yard
```

Examples of this approach are as follows:

```
2.cm+4.inch
=> 4.78740157480315

(24).to_feet
=> 2.0
```

Links to Further Reading

The following list features books and websites that you can use as further reference for both the language Ruby and SketchUp's Ruby API. Keep this list handy as a reference.

For Ruby:

- Clinton, Jason D. *Ruby Phrasebook*. Boston: Pearson Education, 2009.
- Fitzgerald, Michael. *Ruby Pocket Reference*. Sebastopol, CA: O'Reilly Media, 2007.
- www.ruby-lang.org/en—The main website for the Ruby language.

- http://tryruby.org—A well-made interactive site that lets you learn and try out Ruby interactively.

- http://ruby-doc.org—The main documentation for Ruby objects and methods (as of this writing, the Ruby version in SketchUp is 1.8.6). This site also features some introductory texts.

- http://en.wikibooks.org/wiki/Ruby_Programming—An online book on Ruby programming.

- www.techotopia.com/index.php/Ruby_Essentials—Another online book.

- www.zenspider.com/Languages/Ruby/QuickRef.html—A cheat sheet for Ruby.

For SketchUp's API:

- Scarpino, Matthew. *Automatic SketchUp: Creating 3D Models in Ruby.* Walnut Creek, CA: Eclipse Engineering, 2010.

- https://developers.google.com/sketchup—The official documentation of the SketchUp API.

- www.alexschreyer.net/cad/sketchup-ruby-api-cheatsheet—This is a version of the cheat sheet included in Appendix C, which I update whenever a new SketchUp version is released.

- https://groups.google.com/forum/#!forum/google-sketchup-developers—Google's forum for SketchUp developers.

- http://forums.sketchucation.com/viewforum.php?f=180—A forum for Ruby API developers on SketchUcation.

- http://cfcl.com/twiki/bin/view/SketchUp/Cookbook/WebHome—A cookbook-type reference for Ruby code.

Creating Geometry with Ruby

Creating geometry is easy to do with the API. In essence, you add a new entity (a geometry object) to the collection of entities in the current model. To do this, you can use any of the methods of the `Entities` object. For example, `add_line` adds a line based on two points in 3D space. Here is the complete list of available methods and properties of the `Entities` collection:

```
Entities (Parent: Object)

.[] .add_3d_text .add_arc .add_circle
.add_cline .add_cpoint .add_curve
.add_edges .add_face
.add_faces_from_mesh .add_group
.add_image .add_instance .add_line
.add_ngon .add_observer .add_text
.at .clear! .count .each .erase_entities
.fill_from_mesh .intersect_with .length
.model .parent .remove_observer
.transform_by_vectors
.transform_entities
```

In addition to these methods, you can also create a "raw" mesh directly by using the `PolygonMesh` object. Its methods and properties are as follows:

```
PolygonMesh (Parent: Object)

.add_point .add_polygon .count_points
.count_polygons .new .normal_at
.point_at .point_index .points
.polygon_at .polygon_points_at .polygons
.set_point .transform! .uv_at .uvs
```

To be able to add any of these objects to your model, you must first retrieve the `Entities` collection in the current ("active") model. Because we are using the Ruby Code Editor, we can simply use *ent* as a shortcut variable for this collection. Just make sure you leave the three top lines of code in there and add yours below them. This is the default code:

```
mod = Sketchup.active_model # Open model
ent = mod.entities # All entities in model
sel = mod.selection # Current selection
```

Once you have this, simply add to the collection whatever you need.

```
new_line = ent.add_line ([0,0,0],[10,10,10])
```

As you can see here, we are again giving SketchUp 3D points as coordinate arrays. Technically, we have given SketchUp `Point3d` objects here. While the shorthand version shown here works well, the longer version would have been:

```
pt1 = Geom::Point3d.new(0,0,0)
pt2 = Geom::Point3d.new(10,10,10)
new_line = ent.add_line (pt1, pt2)
```

This works similarly when you need to use a vector. As you may have expected, there is also a `Vector3d` object in SketchUp. You can create it in an analogous fashion. Either use the direct triplet entry (which this time represents a direction rather than a 3D point) or create a `Vector3d` object using its `new` method:

```
vec1 = [x,y,z]
vec1 = Geom::Vector3d.new(x,y,z)
```

But enough of the theory for now! Let's take a look at some examples that show us how we can create geometric objects using scripting.

Lots of Boxes

To get started, let's create a 3D grid of regularly spaced boxes. In the following code snippet, we'll define a number of boxes per axis *n*, a box spacing *s*, and a box width *w*. We'll then do three loops: one for multiples along the x axis (the red axis), one for multiples along the y axis (the green axis), and a third for the z axis (the blue axis). We will also use the `pushpull` method of the `Face` object to extrude the faces into boxes.

Paste the following code into the Ruby Code Editor in SketchUp and click on Run Code:

```
# Creates a 3D grid of similar boxes

ent = Sketchup.active_model.entities

n = 5  # number of boxes
s = 100 # spacing
w = 20  # width

(0..n-1).each { |i|
 (0..n-1).each { |j|
  (0..n-1).each { |k|
   # add a face first
   face = ent.add_face [i*s,j*s,k*s],[i*s,j*s+w,k*s],[i*s+w,j*s+w,k*s],
        [i*s+w,j*s,k*s]
   # then pushpull it to get a box
   face.pushpull -w

  }
 }
}
```

The result will look like **Figure 6.8**.

As this demonstrates, creating extruded geometry is very easy. First, we needed to create a face. This was done using the `add_face` method. As always, the face was simply added to the `Entities` collection.

Instead of just adding it, though, we assigned it at the same time to the variable *face*. This gave us a way to apply the `pushpull` method to the face and extrude it by a given amount w.

You might have noticed that w is entered as a negative number (there is a minus sign in front of it). This is because whenever you create a face in SketchUp, the white side of the face (the "front") points downward. A positive w would have extruded it downward as well. Because we want to extrude it upward, w must be negative (or you would have needed to first flip the face's orientation using its `reverse!` method).

Of course, you have to be careful with this script. As you can imagine, changing a single number (n) significantly impacts the amount of geometry created. After all, each cube consists of 18 elements: 6 faces and 12 edges.

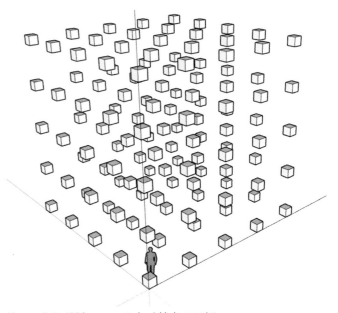

Figure 6.8: 125 boxes created quickly by a script

This gets the job done, but it is not very efficient. We are creating geometry for each of the many similar boxes. Ultimately, all these faces, edges, and vertices need to be stored in the SketchUp file that you save to your hard disk. The next section shows you how to improve on this using components and transformations.

Lots of Boxes with Color

Let's create a slight variation of the preceding code snippet. This time, we'll put each of the boxes into a group, and we'll add color to them all!

Paste the following code into the Ruby Code Editor in SketchUp and click on Run Code:

```
# Creates a 3D grid of colored boxes

ent = Sketchup.active_model.entities

n = 5   # number of boxes
s = 100 # spacing
w = 50  # box width

(0..n-1).each { |i|
 (0..n-1).each { |j|
  (0..n-1).each { |k|
    # create a group for each box
    group = ent.add_group
    # add the face to the group's entities
    face = group.entities.add_face [i*s,j*s,k*s],[i*s,j*s+w,k*s],
          [i*s+w,j*s+w,k*s],[i*s+w,j*s,k*s]
    # add a material (an RGB color)
    face.back_material = [(255/n*i).round,(255/n*j).round,
                          (255/n*k).round]
    # now extrude the box
    face.pushpull -w
  }
 }
}
```

This should at least significantly improve the visuals of the previous example. **Figure 6.9** shows what you will see.

What did we do differently this time? We didn't just add the faces to the `Entities` collection as before; this time, we first added a group to that collection using the `add_group` method. Because we assigned a variable to the group (*group*), we were able to add geometry to the `Entities` collection of that group. That is all you need to create grouped geometry using Ruby.

The other trick we used was to add a material to the face *before* we extruded it. This added the material to the entire box by way of extrusion.

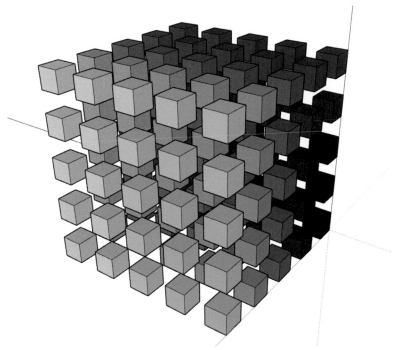

Figure 6.9: Boxes colored in the full spectrum

As you can see, adding a color (a "material") to a face is very easy, too. You just need to use one of the two related methods of the face object: `material` for the front side and `back_material` for the reverse side. As before, the reverse side (the blue side) is pointing in the direction of our extrusion, so we need to apply the material to that side.

Materials can be applied as a reference to a material that is already installed in SketchUp (check for these in the Materials window), as a color name, or—as in our case—as a color defined by RGB (red, green, blue) values. The following are all valid material assignments (assuming that `mat1` is a material that was previously defined in your code):

```
face.material = mat1
face.material = "red"
face.material = 0xff0000
face.material = "#ff0000"
face.material = Sketchup::Color.new(255, 0, 0)
face.material = [1.0, 0.0, 0.0]
face.material = [255, 0, 0]
face.material = 255
```

One thing that you will notice when you create and use a lot of different colors is that SketchUp will have to add all of them to the already defined materials, which might take some time. You can always review the result of your creation in the Materials window under In Model by clicking on the house symbol (see **Figure 6.10**).

TIP

Whenever you define colors using RGB values, keep in mind that the maximum value for each of them is 255. Here are some sample colors:

```
[0,0,0]        # black
[255,0,0]      # red
[255,255,0]    # yellow
[125,125,125]  # a medium gray
```

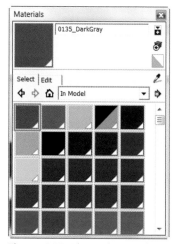

Figure 6.10: The newly created materials

Creating a Small City

Let's try something similar now: Create boxes with a random height—and let's call it a city. As you can see in the following code, we are just producing a 2D grid of boxes. This time, however, we are using feet as the dimension. And instead of giving all of the boxes the same height, we are now using Ruby's `rand` method to randomize their height (between 100 and 200 feet as you can see in **Figure 6.11**).

```
# Creates a small random city

ent = Sketchup.active_model.entities

n = 10    # number of buildings
s = 100.feet # spacing
w = 60.feet  # base width

(0..n-1).each { |i|
  (0..n-1).each { |j|
    # add a face first
```

```
        face = ent.add_face [i*s,j*s,0],[i*s,j*s+w,0],[i*s+w,j*s+w,0],
            [i*s+w,j*s,0]
    # then pushpull it to get a box
    height = 100.feet + rand(100).feet
    face.pushpull -height
  }
}
```

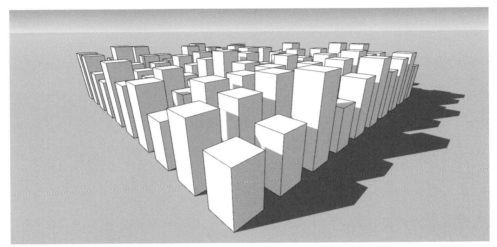

Figure 6.11: A small city with buildings of random heights

Figure 6.11 shows how to use dimensions other than inches in a script and how to employ the built-in random number generator to add variation to your creations. Giving anything generated by a computer at least some randomness usually adds quite a bit of realism. We will use this later with transformations, as well.

Randomizing Extrusions

Instead of applying randomized extrusions to newly created geometry, this example applies them to already existing geometry, which allows us to learn about using selections.

First, create some geometry in your model; you need to have several faces at your disposal. For this example, I used one of SketchUp's preinstalled spheres as well as a square face with intersecting lines (see **Figure 6.12**). Make sure the front (the white side of each face) was always pointing outward and upward; use the Reverse Faces right-click tool if you need to correct any. Then select everything. Paste the following (very short) snippet into the Ruby Code Editor in SketchUp and click on Run Code:

```
# Randomly extrudes selected faces

sel = Sketchup.active_model.selection
```

```
sel.each do |e|
  # First check if this is a face
  if e.typename == "Face"
    # Then extrude it
    e.pushpull rand(100)
  end
end
```

This time, we are not adding anything to the Entities collection (which is why there is no ent in the code). Instead, we are using what we have selected by employing Sketchup. active_model.selection. As shown in **Figure 6.12**, the faces extruded randomly.

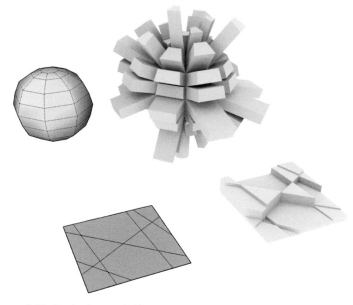

Figure 6.12: Randomly extruded faces

As evident in the code, sometimes it makes sense to check the type of entity you have selected before you do anything to it (thereby preventing errors). This is done using the typename method, and some possible entity types are Face, Edge, ConstructionPoint, ComponentInstance, and Group.

Using Formulas for Patterned Panels

Scripting geometry becomes especially effective (and beautiful) when we use a nonlinear formula to define size, position, color, and so forth. The next piece of code creates a rectangular grid of circles on the x-y plane (the ground) where the circle size is determined in two directions by a sine function.

Paste the following code into the Ruby Code Editor in SketchUp and click on Run Code:

```ruby
# Creates a square panel with a sinusoidal hole pattern

ent = Sketchup.active_model.entities

width = 36
n = 10
s = width/(n+1).to_f

# add the square
ent.add_face [0,0,0],[width,0,0],[width,width,0],[0,width,0]

(0..n-1).each { |i|
  (0..n-1).each { |j|
    # add the circles
    ent.add_circle [s+i*s,s+j*s,0], [0,0,1],
    sin(i/(n-1).to_f*PI)*s/5.0+sin(j/(n-1).to_f*PI)*s/5.0
  }
}
```

After running the script, what you see on the x-y plane should look like **Figure 6.13**.

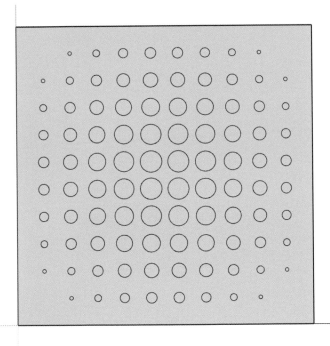

Figure 6.13: Panel layout with sinusoidal pattern

We can now take this geometry as the basis for further modeling. In this case, I used SketchUp's standard Push/Pull tool to extrude the patterned panel (there is no need to do that using code!). When we placed the circles on top of the square face, their faces were automatically separated from the underlying face. This allowed me to use a single push/pull to finish the panel. A rendering could then look like **Figure 6.14**.

Figure 6.14: Rendered panels

You can get creative with this little bit of code if you replace some of the `sin` functions with `cos` or any other mathematical formula. You can also experiment with some of the factors that end up determining circle diameter (the last parameter in the `add_circle` method) or the expression between the brackets in the sine function.

TIP

One thing that may stump you when doing calculations in Ruby is Ruby's reluctance to give a floating point value after dividing two integers. This is the reason you see `to_f` in a few places in the code. This little snippet converts any number to a "float," a real number. Alternatively, you could initialize the variables with an extra digit, as in `width = 36.0`.

Plotting Data from Text Files

When you create geometry, parameters might need to come from a source outside of SketchUp. This would be the case if you had an Excel spreadsheet with `x,y,z` coordinates (as would be available from a LIDAR laser scan or from a land surveyor). You might also have other software (structural or other building analysis software) export parameters that you could use for geometry creation. In any case, let's assume for this example that you had point coordinates stored in a text file in the following format:

```
81.82821,16.42147,205.31
83.07821,16.33147,204.91
83.57821,10.31147,204.87
82.20821,10.43147,204.50
```

As you can see, each point is printed on a single line in the x,y,z format. Units are feet in this case, but we can ignore them for now (which defaults the units to inches, of course).

Let's explore how to import and plot them in your SketchUp model. First, we have to read from the text file line by line and then plot each data point to SketchUp. To do this, we will add construction points (little tick marks that you usually create using the Tape Measure tool) with the add_cpoint method.

First, paste the point data listed previously, or any similarly formatted data, into a text file (file naming isn't crucial). Then paste the following code snippet into the Ruby Code Editor and hit Run:

```
# This code asks for a file, loads 3D point data and
# draws a construction point for each imported point.
# Data must be in the format x,y,z per line in your file

ent = Sketchup.active_model.entities

# Starts at line zero, modify if you have a header
startline = 0

# Use the openpanel to get the file name
filename = UI.openpanel 'Select Data File'

# Iterate through all lines in the file and plot the cpoints
if filename != nil
  file = File.open(filename)
  file.each { |line|
    if file.lineno >= startline
      pointXYZ = line.strip.split(",")
      ent.add_cpoint [pointXYZ[0].to_f,pointXYZ[1].to_f,pointXYZ[2].to_f]
    end
  }
  file.close
end
```

This code first asks you to provide the filename by showing you the openpanel dialog. It then opens the file and iterates through each line. Once it reads each line, it applies two methods to the line object: First, it uses strip to remove any white spaces around your data. Then it uses split(",") to split each line into an array of elements. As you can see, you can conveniently supply the character that is used to make the split (a comma, in my case). Finally, the x,y,z data is used to add construction points to the Entities collection (and therefore your model).

This little example demonstrated a few new things. You learned how to work with the openpanel (as you might have guessed, there is also a savepanel in SketchUp) and then use the supplied file reference. You also learned how to work with external files and the very useful split method.

Figure 6.15 shows the result of importing a large file consisting of point coordinates using the method described here. The raw data in this case came from a LIDAR flyover of a historic church.

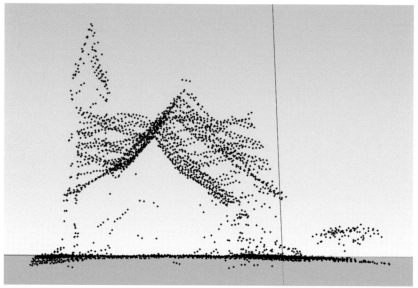

Figure 6.15: Construction points from an imported file

Working with Text Files

As the preceding example shows, it is quite easy to work with text files. The common approach to reading from a file is this:

```
filename = 'data.txt'
f = File.open(filename)
f.each { |line|
  # Do something with the line here
}
f.close
```

Alternatively, you could replace the middle three lines of this code with `content = f.readlines`, which reads the entire file as an array into the `content` variable. You would then use `content.each` to iterate through the lines.

To write to a file, use this syntax:

```
filename = 'data.txt'
f = File.new(filename,'w')
```

```
f.puts 'Hello world!'
f.close
```

As you might have guessed, the `'w'` at the end of the second line prepares the file for writing. Another useful switch is `'a'`, which appends anything you write to the file at its end (instead of overwriting any existing content).

Saving Vertices to a Text File

In this example, let's reverse the process of the last one. Assume you have created a very complex shape in SketchUp and now need to export edge coordinates (vertices) to a text file, so that you can use them in different software (in Excel, for example). The following code takes your current selection of objects and writes a text file with x,y,z data for each unique vertex.

Create some geometry (that has edges) in SketchUp, select it (make sure it is not grouped), paste the following code snippet into the Ruby Code Editor, and hit Run:

```
# Write all vertices in selection to file

# Current selection
sel = Sketchup.active_model.selection

# Empty array for vertices
verts = []

# Iterate over the selection and store vertices
sel.each do |e|
 # Only consider edges
 if e.typename == "Edge"
  # Add vertices to array
  verts.push e.vertices
 end
end

# Remove any sub-arrays (if necessary)
verts.flatten!
# Remove duplicates in array
verts.uniq!

# Ask for export file
filename = UI.openpanel 'Text file for vertex coordinates'
if filename
  f = File.open(filename, 'w')
  # Write x,y,z coordinates in default units (inches)
  verts.each { |v|
```

```
        f.print v.position.x.to_inch,',',v.position.y.to_inch,',',
                v.position.z.to_inch,"\n"
      }
    f.close
end
```

As you can see, we are again using the `Selection` collection of objects. In a first step, we iterate over this selection using `sel.each`. As in the earlier example, we check that we are dealing with an edge and then add its (two) vertices to the `verts` array using the `push` method.

Obviously, some of those vertices will be duplicates. Consider that two edges of a square meeting at one point will write this point twice to the collection. To fix this, we use the Ruby array method `uniq!`, which removes duplicates from the array.

Finally, we print all of the x,y,z coordinates to a new text file, which the user can select through an `openpanel`. This is done by first getting the position of each vertex, then getting its x,y,z value, and, finally, converting these values to SketchUp's default units using the `to_inch` method (otherwise, they will be stored in your localized unit system—e.g., as *8'4"*). You can now check that everything worked correctly by using the code in the "Plotting Data from Text Files" section to read the points back into SketchUp. (See **Figure 6.16**.)

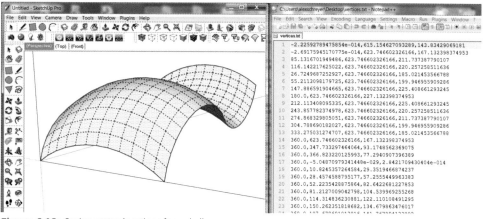

Figure 6.16: Saving vertex locations for a shell

Transformations Change Things Up

What you already know in SketchUp as the Move, Rotate, and Scale tools is technically called a *transformation*. Therefore, you actually won't find a move or rotate method in SketchUp's Ruby API. While this might be less obvious or convenient, using transformations to create copies of objects and/or move them is not too hard with Ruby.

A transformation is based on its equivalently named object. The following snippet lists all of the available methods.

```
Transformation (Parent: Object)

.* .axes .clone .identity?
```

```
.interpolate .inverse .invert!
.new .origin .rotation .scaling
.set! .to_a .translation
.xaxis .yaxis .zaxis
```

Although this (and the online documentation) might appear a bit confusing, let's look at just a subset of the available functionality. This provides enough functionality for what we want to do. The relevant methods are these:

```ruby
# Point Transformation—use for placing objects
t1 = Geom::Transformation.new( point )

# Translation Transformation—use for moving
t2 = Geom::Transformation.translation( vector )

# Rotation Transformation—use for rotating
t3 = Geom::Transformation.rotation( center, axis, angle )

# Scaling Transformation—use for scaling
t4 = Geom::Transformation.scaling( center, scale )
t4 = Geom::Transformation.scaling( center, xscale, yscale, zscale )
```

As you can see, they all use perfectly understandable parameters. If you want to move something from location A to location B, then the vector between those points (expressed as "go *m* units in the x direction, *n* units in the y direction, and *o* units in the z direction") is enough information to get it there. And if you want to rotate something, SketchUp needs to know *about which point* you want it to rotate, *about which axis* you want it to rotate, and *by how many degrees* it should turn. Finally, the scaling translation accepts either a uniform scale or different scale factors in each coordinate direction.

In order to use these transformations, you need to add them as parameters to appropriate methods. There are various approaches to this: Use `transform_entities` on the model's `Entities` collection as a universal approach. You can also directly apply a transformation to groups or components, especially when you place an instance of a component (the transformation describes the insertion point in that case).

Let's look at this in action in a short code snippet. Select an object in your model (if you select multiple, only the first gets used). Then paste the following code into the Ruby Code Editor and click Run Code:

```ruby
# Test transformations on selected object

mod = Sketchup.active_model # Open model
ent = mod.entities # All entities in model
sel = mod.selection # Current selection

# Get the center of the selected object
center = sel[0].bounds.center
```

259

```
# Define all possible translations
t1 = Geom::Transformation.new([100,0,0])
t2 = Geom::Transformation.translation([100,0,0])
t3 = Geom::Transformation.rotation(center, [1,0,0], 45.degrees)
t4 = Geom::Transformation.scaling(center, 2)

# Choose one of them—for now
t = t2

# Transform one object
ent.transform_entities(t, sel[0])
```

You can test the main transformations simply by modifying the t = t2 statement. Feel free to experiment with the code a bit to test it out.

If you want to apply more than one transformation, then replace this line with something like t = t2 * t3 * t4 (i.e., multiply the ones you want to apply). This combines t2, t3, and t4 into a single transformation.

The following examples apply these principles in various ways.

Lots of Boxes Using Components

This is essentially the same example that we did earlier. In SketchUp, however, the approach that takes less computational effort and less space on the hard disk than just creating lots of geometry is usually to create one geometry item (the box, in this case), convert it to a component, and then just insert component instances. Instead of saving a whole lot of geometry, only insertion points are thus being stored. This results in smaller file sizes and models that handle much better (during zooming, orbiting, etc.).

The following code does this by creating a grouped box, converting it into a component, and then copying (placing) it multiple times using transformations.

```
# Creates lots of boxes using components

ent = Sketchup.active_model.entities

s = 100
w = 10
n = 10

group = ent.add_group
face = group.entities.add_face [0,0,0],[w,0,0],[w,w,0],[0,w,0]
face.pushpull -w
comp = group.to_component

(0..n).each { |i|
 (0..n).each { |j|
  (0..n).each { |k|
```

```
      transformation = Geom::Transformation.new([i*s,j*s,k*s])
      ent.add_instance(comp.definition, transformation)
    }
  }
}
```

As you can see, this code creates a grid of boxes similar to **Figure 6.8**. In contrast to what we did before, this time we created *one* box that has the appropriate geometry. Then we changed it into a component using the `group.to_component` method. It is only after this that we loop through the locations, define an insertion point using a transformation, and add a component instance (as `comp.definition`) using the `add_instance` method at that insertion point.

Building a Curved Wall

Let's expand a bit on the previous code. In this snippet, we'll take a preexisting component—a 1' × 1' × 2' block in the example used here—and place it multiple times while rotating it. The rotation is again guided by a combination of sine and cosine in the horizontal and vertical directions.

To use this code, you must first have a component instance in your model (the block you see in **Figure 6.17** is the one used here; place it anywhere you like). Before you run the code, make sure you select the component—otherwise, nothing happens.

```
# Rotate-copy placement of a component

mod = Sketchup.active_model # Open model
ent = mod.entities # All entities in model
sel = mod.selection # Current selection

s = 28 # x-spacing
t = 12 # z-spacing
h = 20 # number in x
v = 20 # number in z
max_deg = 90 # Max "opening" angle

if (sel[0].typename == 'ComponentInstance')
  v.times { |j|
    h.times { |i|

      # Place a brick at the specified location
      t1 = Geom::Transformation.translation([i*s,0,j*t])
      instance = ent.add_instance(sel[0].definition, t1)

      # Calculate angle and rotate the brick vertically
      angle = (sin((i*180/h).degrees)+sin((j*360/v).degrees))*max_deg/2
```

```
    t2 = Geom::Transformation.rotation instance.bounds.center,
      [0,0,1],
      angle.degrees
    instance.transform! t2

    }
  }
end
```

There are a few more parameters in this code than we had before (brick spacing in both directions, number of bricks, and opening angle), but this gives you the flexibility to experiment with them and generate variations in the results.

In the code, we first check that the selected item is actually a `ComponentInstance`. Next, we iterate through each horizontal and vertical brick placement. In this iteration, we use a `translation` transformation to place the brick and then we apply a rotation to the new component instance. This allows us to rotate the brick "in place."

Figure 6.18 shows a rendering of a generated wall.

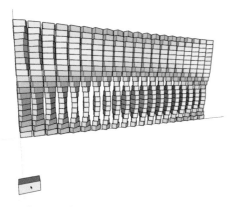

Figure 6.17: Single selected brick and generated wall   **Figure 6.18:** Rendering of generated wall

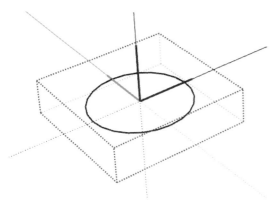

Placing Components on Faces

In this example, we take a component (a simple circle) and copy it onto the center of each selected face of some existing geometry. To make the code work, prepare the component first: Create a circle and turn it into a component using SketchUp's standard tools. Edit the component so that the circle is centered on the component's coordinate system, as shown here.

Create any shape you like that has faces. In my example, I created a shell using the Extrude Edges by Edges tool from the Extrusion Tools plugin (see Chapter 4). **Figure 6.19** shows both the single component and all ungrouped faces selected and ready for the code to do its magic.

Whenever you work with components in Ruby, it is very important to have the component's axes (the ones that you see when you edit or double-click the component) in the correct location. Copies will be placed using a component's origin. Double-check their placement if your results don't look as expected.

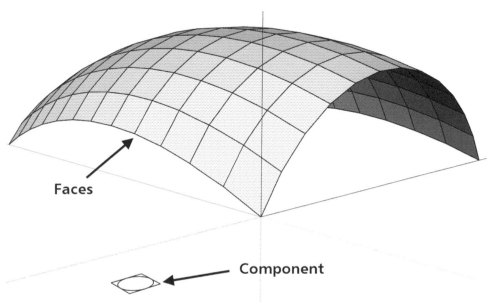

Faces

Component

Figure 6.19: Selection for example

```
# Use this to map a component onto the center of all selected faces
mod = Sketchup.active_model # Open model
ent = mod.entities # All entities in model
sel = mod.selection # Current selection

# Get the one component's definition
comp = nil
sel.each { |e|
  if e.typename == "ComponentInstance"
    comp = e.definition
  end
}

# Iterate through all ungrouped faces
sel.each { |e|
  if e.typename == "Face"
```

```
  # Place a copy at the center of the bounding box
  center = e.bounds.center
  t = Geom::Transformation.new (center, e.normal)
  new = ent.add_instance (comp, t)

  # Now scale objects by some criteria
  new.transform! Geom::Transformation.scaling center,
   center.z/100

  # Explode it so that we can remove the face afterwards
  new.explode

 end
}
```

When you run this code, it first looks through your selection and isolates the one ComponentInstance (the circle). Then it iterates through each face and places a new instance at the center of the face. It does this by finding the center point using the bounds.center method of the Entity object (which gives the center of the bounding box). The new instance is then placed with its orientation (its "normal" vector) aligned with the face's normal vector (which describes the direction perpendicular to the face).

A scaling transformation is applied to the circle, which varies the circle's scale by how far away from the ground it is. This results in small circles ("openings") at the base of the shell and the largest ones at the apex. As a last step, the component instance is exploded, which lets us remove the circle's faces manually, leaving nice circular openings in the shell. The result is shown in **Figure 6.20**.

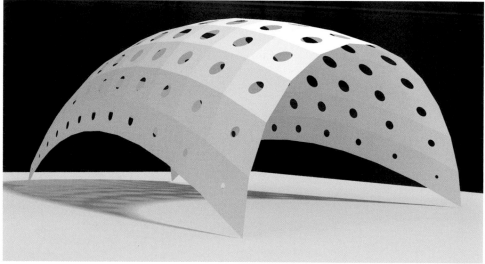

Figure 6.20: Rendered view of canopy after removal of circle faces

Feel free to experiment with different components using this code. You can even place ones that have thickness (e.g., a cylinder) using this method.

Randomizing Everything

In the following code, we revisit the use of random numbers and combine this with what we learned about transformations. Specifically, we take the selected objects (all must be component instances) and apply a random rotation as well as random scaling to each. Before you run this code, add some components to your model—trees or other shrubbery make for great examples. Just make sure you are not using face-me components.

```ruby
# Apply randomized rotation and scaling to selection

mod = Sketchup.active_model # Open model
ent = mod.entities # All entities in model
sel = mod.selection # Current selection

max_rotation_angle = 90
size_var = 0.5 # Keep below 1

sel.each { |e|

  if e.typename == "ComponentInstance"

    # Get the center of the selected object
    center = e.bounds.center
    # Also get the base center for scaling
    base = center.clone
    z_height = (e.bounds.max.z—e.bounds.min.z)
    base.z = base.z—z_height/2

    # Transform this object
    t1 = Geom::Transformation.rotation(center, [0,0,1],
      (rand * max_rotation_angle).degrees)
    t2 = Geom::Transformation.scaling base,
      1—size_var/2 + rand*size_var
    # Combine transformations
    t = t1 * t2

    e.transform! t

  end

}
```

As you can see in **Figure 6.21**, this little snippet can add quite a bit of realism, especially to natural objects like trees.

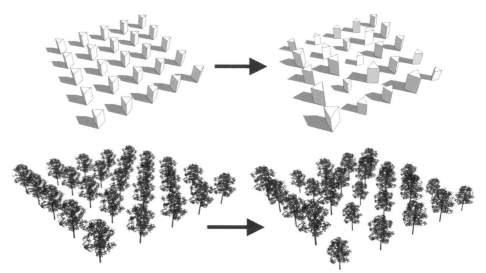

Figure 6.21: Randomized rotation and scaling (left: before; right: after)

To create this effect, the code first takes the entire `Selection` collection and iterates through each element using `sel.each`. It then checks that the current element is a `ComponentInstance`. You could apply this to all kinds of entities; however, `transform!` is a method that works only for the `ComponentInstance` object. You would have to modify the transformation slightly for other objects.

Next, the code finds the center of the bounding box of the component as well as the center of its base. To find the base center, we needed to calculate the height of the box and then subtract half the height from the z value.

Finally, we define two transformations, one for rotation and one for scaling, and combine them by multiplication. As you can see in the code, two variables allow you to modify the maximum rotation angle as well as the maximum size variation (as a ratio). Experiment with these values until you get the results you are after.

Attracted to Attractors

The following examples show how you can modify geometry based on proximity to some other object (or any relationship you can think of, such as "height from ground"). This often yields very interesting designs and lets you express relationships visually in your designs.

At the minimum, what is required to make these examples work is (1) some geometry that you intend to modify and (2) one or more objects that act as attractors, on which the modification is based. A good example of the first is a facade panel whose color or shape you intend to modify. It doesn't have to be a face, though—you could instead use a group or a component instance.

The attractor can be any object. For convenience, I like using either component instances (the default figure in SketchUp's template—lately, "Susan"—for example) or groups or construction points, but it is up to you to decide what works best in your case. Just modify the code examples slightly to make them work if you deviate from the examples.

Coloring Faces by Proximity

This code takes several faces and one or more component instances and then colors the faces according to their proximity to the centers of the component instances (the "attractors"). A face that is close to an attractor gets a blue color, one that is far away turns yellow, and any faces in between receive colors from a graduated spectrum.

This example uses a facade made up of several 4′ × 8′ glass panels. These are simply faces (rectangles) copied edge to edge. We leave them all ungrouped so that the code can identify them as faces. In addition, we place one to three copies of the default SketchUp "person" in various locations and with differing distances to the facade (the closest is the one on the left side). Before running the code, select all of the faces and the people components. **Figure 6.22** shows the raw setup with the facade panels and only one attractor. It also illustrates the result of running the code on this setup.

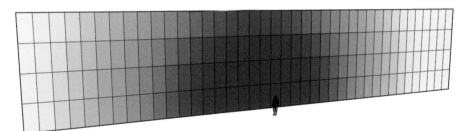

Figure 6.22: Single-attractor-based coloring of facade panels

Paste this code into the Ruby Code Editor, select the necessary objects in your SketchUp model, and then click on Run:

```
# Color faces by proximity to component attractors

sel = Sketchup.active_model.selection

# Array with attractor center points
attractors = []

# Adjust this value to play with the scaling
max_val = 500

# Get all ComponentInstances as attractor points
sel.each { |e|
```

```
    if e.typename == 'ComponentInstance'
      attractors.push e.bounds.center
    end
}

# Iterate through objects and scale based on distance
if attractors.length > 0

  # Do this for each face
  sel.each { |e|
    if e.typename == "Face"

      # Get the center of the face
      center = e.bounds.center

      # Calculate minimum distance between face and
      # all attractors
      dist = 1000000 # Start with a large number

      # Find the smallest distance
      attractors.each { |apoint|
        dist_calc = apoint.distance center
        dist = dist_calc if dist_calc < dist
      }

      # Adjust color of face between blue and yellow
      e.material = [(dist/max_val*255).round,(dist/max_val*255).round,
      (255-dist/max_val*255).round]

    end
  }

end
```

This results in a very colorful and interesting facade pattern. **Figure 6.23** shows how this would look in a rendering of a building.

Similar to what we did with the previous code, we start this one by finding the component instances and adding their center points to the `attractors` array. After a check that we actually have attractors in our selection (`attractors.length > 0`), we proceed to iterate through the entire selection again—this time, considering only the faces. For each face, we then calculate the distances between the center of the face and each attractor and keep the shortest distance as the basis for our coloring. Finally, we apply materials (colors) to the front faces (making sure they are in the correct orientation this time) by specifying red, green, and blue values.

This code can be modified in many ways to generate different results. At the very least, you can modify `max_val`, which affects the magnitude of the attractors' influence on the coloring. You could also modify the color assignment by using less dramatic color combinations.

Figure 6.23: Rendering of an entire building with attractor-based facade pattern

Scaling Objects by Proximity

Instead of coloring faces according to their proximity to an attractor, you could instead modify their geometry, for example, by scaling them. In this example, we take a wall that is made up of 4′ × 4′ slightly "bulging" wall panels and let the attractors adjust the panels' thickness.

The panels for this example are created from four edge lines using the Soap Skin & Bubble plugin. The component axis is placed at the back of the component definition, where the panels would be attached to a wall. (See **Figure 6.24**.)

This code is very similar to the previous code, the difference being that we now use groups as attractors (two small boxes) and component instances as the objects that we modify. The only major change in the code is at the end, where we apply the scaling transformation. See **Figure 6.25** for the resulting paneled wall.

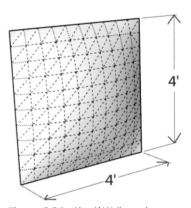

Figure 6.24: 4′ × 4′ Wall panel component

Paste this code into the Ruby Code Editor, select the necessary objects in your SketchUp model, and then click on Run:

```
# Scale groups by proximity to component attractors

sel = Sketchup.active_model.selection

# Array with attractor center points
attractors = []

# Adjust this value to play with the scaling
```

```
max_val = 200

# Get all attractor points
sel.each { |e|
  if e.typename == 'Group'
    attractors.push e.bounds.center
  end
}

# Iterate through objects and scale based on distance
if attractors.length > 0

  # Do this for each group
  sel.each { |e|
    if e.typename == "ComponentInstance"

      # Get the center of the group
      center = e.bounds.center
      # Set the y-value to zero so that we
      # scale "outward"
      center.y=0

      # Calculate minimum distance between face and
      # all attractors
      dist = 1000000 # Start with a large number

      # Find the smallest distance
      attractors.each { |apoint|
        dist_calc = apoint.distance center
        dist = dist_calc if dist_calc < dist
      }

      # Calculate scale
      scale = (dist/max_val)**2
      t = Geom::Transformation.scaling (center, 1, scale, 1)
      e.transform! t

    end
  }

end
```

As you can see in the last few lines of code, this scaling transformation uses the syntax that allows for different scale factors in each dimension. Because the panels are oriented with the y direction (the components' green axis) perpendicular to the wall, all other scale factors remain as 1.

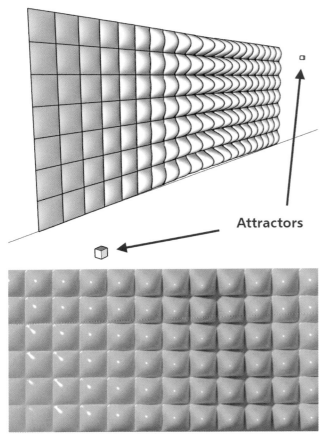

Figure 6.25: Modified panels and the two attractor boxes

Solar-Responsive Design

This script does not use attractors per se. It is, however, an example of how you could use a direction (expressed as a vector) to drive your design. To make our design more responsive to lighting, we'll use the "south" direction in our model to modify the shape of components that we place on a curved surface.

To accomplish this, we'll reuse the code from the *Placing Components on Faces* script from earlier. All we need to do this time is modify the section that transforms the copied `ComponentInstance`.

You can try this if you have various faces in your selection as well as one component. The example shown in **Figure 6.26**, uses a curved surface that was subdivided into many faces and the same circle component as was used with the earlier code. Once you have these, select them, copy the following code into the Ruby Code Editor, and click on Run.

```
# Maps a component onto the center of all selected faces and
# scales based on south orientation
```

```
mod = Sketchup.active_model # Open model
ent = mod.entities # All entities in model
sel = mod.selection # Current selection

# Get the one component's definition
comp = nil
sel.each { |e|
  if e.typename == "ComponentInstance"
    comp = e.definition
  end
}

# Iterate through all ungrouped faces
sel.each { |e|
  if e.typename == "Face"

    # Place a copy at the center of the bounding box
    center = e.bounds.center
    t = Geom::Transformation.new (center, e.normal)
    new = ent.add_instance (comp, t)

    # Now scale objects by south orientation
    scale = 1-[(e.normal.angle_between [0,-1,0]),0.2].max
    new.transform! Geom::Transformation.scaling center, scale, scale,
                   scale*2

    # Explode it so that we can remove the face afterwards
    new.explode

  end
}
```

As you can see in **Figure 6.26**, this code places scaled circles (due to unequal scaling, they actually become ellipses) onto each of the selected faces. If the face's orientation points south, the circles become enlarged, yet the farther the face deviates from this direction, the smaller the circle becomes.

In the preceding code, you can see that we kept the first part as before: We loop through the entities in the selection to find the one ComponentInstance. Next, we again iterate through all the faces and place copies of the circle at their centers (oriented with the face normal). Finally, we scale them based on orientation. To do this, we need to calculate the angle between the face normal and the south orientation. This is conveniently done using the angle_between method, which operates on the face normal and checks it against a unit vector in the south direction ([0,-1,0]). This yields the angle in radians, but since this number is between 0 and 1, it is perfect for us to use as a scale factor. To ensure that the circles don't get too small, we'll pick as a scale factor the maximum value of either the angle or 0.2 using the max method (`[angle, 0.2].max`), which looks for the largest value in an array. Finally, we apply the transformation with a doubled value in the z direction, which creates the ellipses.

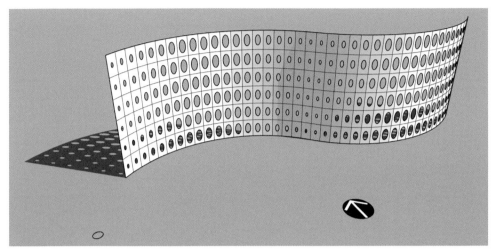

Figure 6.26: Solar-oriented opening placement

What Else Is Possible with This?

As the preceding examples demonstrated, you can create interesting and intriguing geometry using scripting in SketchUp. The combination of manual modeling and Ruby scripting is a very efficient and effective approach to get results fast. If this was your introduction to computational geometry, then feel free to explore further. Start by modifying the code examples and see where this takes you. It is also a good idea to browse the Ruby API documentation for methods that could be of value for your design tasks. Here are some ideas:

- Insert images using the `Image` object.
- Influence the current view using the `Camera` object.
- Geo-locate your model and add geometry using the `LatLong` methods.
- Modify the layers in your model using the methods of the `Layers` collection.
- Add view tabs to your model using the `Pages` collection's methods.
- Add an attribute (a text-based parameter) to an object (e.g., a component) and read it out to a text file. Relevant methods are: `model.set_attribute` and `model.get_attribute`.
- Add a menu item to SketchUp and assign your code to it. Relevant methods are: `menu.add_item` and `menu.add_submenu`.
- Create a `WebDialog` and display some HTML pages in it. You can even make Ruby interact with a WebDialog's content or JavaScript.

If you decide that Ruby scripting is for you, then you might want to look at some of the web links mentioned previously and even explore plugin development. If you created a solution to a problem using Ruby code, then it likely will be useful for someone else, too. Packaging code as a plugin requires only a few more lines of code and is a great way to share your work.

Some Pitfalls and Things to Watch Out For

So SketchUp has crashed after you hit Run in the Code Editor? Here are some things to watch out for:

- **Save both your Ruby code and your SketchUp model early and often!** This will prevent headaches later. It's especially important to save just before you decide to run the code.

- **Errors are okay!** Part of scripting anything is generating (and then fixing) errors. You may observe your code not doing anything at some point. If you are lucky, it will give you an error that sounds meaningful. But often it won't.

  Try running your code again with SketchUp's Ruby Console open. Often, that reveals more information. If that doesn't help, comment out code parts that might be problematic. If the code runs fine without those parts, then you are one step closer to finding the problem.

- **Your code can create geometry at an alarming pace!** If you are writing loops to create geometry, test them with only a few elements first. You can easily grind SketchUp or even your computer to a complete halt. Usually, "killing" SketchUp and restarting it gets you back where you were before (as long as you saved everything, of course).

- **Be careful if your code creates overlapping geometry!** As always in SketchUp, when ungrouped geometry overlaps, SketchUp tries to attach faces or resolve intersections. This can take some time if there are a lot of intersections.

Your Turn!

The following tasks allow you to practice the topics in this chapter:

1. **Design a Panel**
   Create a panel design of your own. Use a 4′ × 4′ panel and populate it with a scripted pattern of boxes, circles, or lines. Create an interesting and aesthetically pleasing pattern.

2. **Panels for CNC Manufacturing**
   Create a panel design that is to be manufactured using a CNC router. Assume a drill bit size and modify your code so that all minimum circle diameters conform to this size.

3. **Create a Parametric Brick Wall**
   Create a brick wall where bricks are offset from course to course (the way they actually would be built). Use the script sample from earlier in this chapter as a basis. Do this first for a straight wall. Then create a curved wall based on parameters of your choosing.

4. **Solar Facade Design**
   Create a solar-responsive design for a facade, a wall, or a screen. Place shading elements on wall panels and size them according to solar orientation.

Appendix A
SketchUp Quick Reference Cards

SketchUp Quick Reference Card

Large Tool Set

Select (Spacebar)		Make Component
Paint Bucket (B)		Eraser (E)
Rectangle (R)		Line (L)
Circle (C)		Arc (A)
Polygon		Freehand
Move (M)		Push/Pull (P)
Rotate (Q)		Follow Me
Scale (S)		Offset (F)
Tape Measure (T)		Dimensions
Protractor		Text
Axes		3D Text
Orbit (O)		Pan (H)
Zoom (Z)		Zoom Extents
Previous		Next
Position Camera		Look Around
Walk		Section Plane

Solid Tools

Outer Shell		Split (Pro)
Intersect (Pro)		Union (Pro)
Subtract (Pro)		Trim (Pro)

Dynamic Components

Interact		Component Options
Component Attributes		

Sandbox (Terrain)

From Contours		From Scratch
Smoove		Stamp
Drape		Add Detail
Flip Edge		

Standard Views

Iso		Top
Front		Right
Back		Left

Style

X-Ray		Back Edges
Wireframe		Hidden Line
Shaded		Shaded with Textures
Monochrome		

Google

Add New Building...		Add Location
Show Terrain		Photo Textures
Preview Model in Google Earth		Share Component...
Get Models...		Share Model...

Display additional toolbars by choosing View > Toolbars from the menu bar.

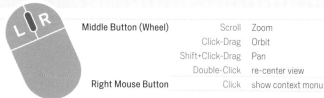

Middle Button (Wheel)	Scroll	Zoom
	Click-Drag	Orbit
	Shift+Click-Drag	Pan
	Double-Click	re-center view
Right Mouse Button	Click	show context menu

Windows

Tool	Operation	Instructions
Arc (A)	Bulge	specify bulge amount by typing a number and Enter
	Radius	specify radius by typing a number, the R key, and Enter
	Segments	specify number of segments by typing a number, the S key, and Enter
Circle (C)	Shift	lock in current plane
	Radius	specify radius by typing a number and Enter
	Segments	specify number of segments by typing a number, the S key, and Enter
Eraser (E)	Ctrl	soften/smooth (use on edges to make adjacent faces appear curved)
	Shift	hide
	Ctrl+Shift	unsoften/unsmooth
Follow Me	Alt	use face perimeter as extrusion path
	Better Way	first Select path, then choose the Follow Me tool, then click on the face to extrude
Line (L)	Shift	lock in current inference direction
	Arrows	up or down arrow to lock in blue direction; right to lock in red; left to lock in green
	Length	specify length by typing a number and Enter
Look Around	Eye Height	specify eye height by typing a number and Enter
Move (M)	Ctrl	move a copy
	Shift	hold down to lock in current inference direction
	Alt	auto-fold (allow move even if it means adding extra edges and faces)
	Arrows	up or down arrow to lock in blue direction; right to lock in red; left to lock in green
	Distance	specify move distance by typing a number and Enter
	External Array	n copies in a row: move first copy, type a number, the X key, and Enter
	Internal Array	n copies in between: move first copy, type a number, the / key, and Enter
Offset (F)	Double-Click	apply last offset amount to this face
	Distance	specify an offset distance by typing a number and Enter
Orbit (O)	Ctrl	hold down to disable "gravity-weighted" orbiting
	Shift	hold down to activate Pan tool
Paint Bucket (B)	Ctrl	paint all matching adjacent faces
	Shift	paint all matching faces in the model
	Ctrl+Shift	paint all matching faces on the same object
	Alt	hold down to sample material
Push/Pull (P)	Ctrl	push/pull a copy of the face (leaving the original face in place)
	Double-Click	apply last push/pull amount to this face
	Distance	specify a push/pull amount by typing a number and Enter
Rectangle (R)	Dimensions	specify dimensions by typing length, width and Enter ie. 20,40
Rotate (Q)	Ctrl	rotate a copy
	Angle	specify an angle by typing a number and Enter
	Slope	specify an angle as a slope by typing a rise, a colon (:), a run, and Enter ie. 3:12
Scale (S)	Ctrl	hold down to scale about center
	Shift	hold down to scale uniformly (don't distort)
	Amount	specify a scale factor by typing a number and Enter ie. 1.5 = 150%
	Length	specify a scale length by typing a number, a unit type, and Enter ie. 10m
Select (Spacebar)	Ctrl	add to selection
	Shift	add/subtract from selection
	Ctrl+Shift	subtract from selection
Tape Measure (T)	Ctrl	create a new Guide
	Arrows	up or down arrow to lock in blue direction; right to lock in red; left to lock in green
	Resize	resize model: measure a distance, type intended size, and Enter
Zoom (Z)	Shift	hold down and click-drag mouse to change Field of View

© 2010 Google Inc.

(Used by permission of Google Inc.)

277

SketchUp Quick Reference Card

Large Tool Set

Select (Spacebar)		Make Component
Paint Bucket (B)		Eraser (E)
Rectangle (R)		Line (L)
Circle (C)		Arc (A)
Polygon		Freehand
Move (M)		Push/Pull (P)
Rotate (Q)		Follow Me
Scale (S)		Offset (F)
Tape Measure (T)		Dimensions
Protractor		Text
Axes		3D Text
Orbit (O)		Pan (H)
Zoom (Z)		Zoom Window
Zoom Extents		Previous
Position Camera		Walk
Look Around		Section Plane

Solid Tools

Outer Shell		Intersect (Pro)
Union (Pro)		Subtract (Pro)
Trim (Pro)		Split (Pro)

Sandbox (Terrain)

From Contours	
From Scratch	
Smoove	
Stamp	
Drape	
Add Detail	
Flip Edge	

Google

Add Location...	
Toggle Terrain	
Add New Building...	
Photo Textures	
Preview Model in Google Earth	
Get Models...	
Share Model...	
Share Component...	

Dynamic Components

Interact	
Component Options	
Component Attributes	

Style

X-Ray	
Back Ege	
Wireframe	
Hidden Line	
Shaded	
Shaded with Textures	
Monochrome	

Standard Views

Iso	
Top	
Front	
Left	
Right	
Back	

To add other tools, right-click the toolbar (at the top of your document window) and choose "Customize Toobar..."

Middle Button (Wheel)	Scroll	Zoom	
	Click-Drag	Orbit	
	Shift+Click-Drag	Pan	
	Double-Click	re-center view	
Right Mouse Button	Click	show context menu	

Mac OS X

Tool	Operation	Instructions
Arc (A)	Bulge	specify bulge amount by typing a number and Enter
	Radius	specify radius by typing a number, the R key, and Enter
	Segments	specify number of segments by typing a number, the S key, and Enter
Circle (C)	Shift	lock in current plane
	Radius	specify radius by typing a number and Enter
	Segments	specify number of segments by typing a number, the S key, and Enter
Eraser (E)	Option	soften/smooth (use on edges to make adjacent faces appear curved)
	Shift	hide
	Option+Shift	unsoften/unsmooth
Follow Me	Command	use face perimeter as extrusion path
	Better Way	first Select path, then choose the Follow Me tool, then click on the face to extrude
Line (L)	Shift	lock in current inference direction
	Arrows	up or down arrow to lock in blue direction; right to lock in red; left to lock in green
	Length	specify length by typing a number and Enter
Look Around	Eye Height	specify eye height by typing a number and Enter
Move (M)	Option	move a copy
	Shift	hold down to lock in current inference direction
	Command	auto-fold (allow move even if it means adding extra edges and faces)
	Arrows	up or down arrow to lock in blue direction; right to lock in red; left to lock in green
	Distance	specify move distance by typing a number and Enter
	External Array	n copies in a row: move first copy, type a number, the X key, and Enter
	Internal Array	n copies in between: move first copy, type a number, the / key, and Enter
Offset (F)	Double-Click	apply last offset amount to this face
	Distance	specify an offset distance by typing a number and Enter
Orbit (O)	Option	hold down to disable "gravity-weighted" orbiting
	Shift	hold down to activate Pan tool
Paint Bucket (B)	Option	paint all matching adjacent faces
	Shift	paint all matching faces in the model
	Option+Shift	paint all matching faces on the same object
	Command	hold down to sample material
Push/Pull (P)	Option	push/pull a copy of the face (leaving the original face in place)
	Double-Click	apply last push/pull amount to this face
	Distance	specify a push/pull amount by typing a number and Enter
Rectangle (R)	Dimensions	specify dimensions by typing length, width and Enter ie. 20,40
Rotate (Q)	Option	rotate a copy
	Angle	specify an angle by typing a number and Enter
	Slope	specify an angle as a slope by typing a rise, a colon (:), a run, and Enter ie. 3:12
Scale (S)	Option	hold down to scale about center
	Shift	hold down to scale uniformly (don't distort)
	Amount	specify a scale factor by typing a number and Enter ie. 1.5 = 150%
	Length	specify a scale length by typing a number, a unit type, and Enter ie. 10m
Select (Spacebar)	Option	add to selection
	Shift	add/subtract from selection
	Option+Shift	subtract from selection
Tape Measure (T)	Option	create a new Guide
	Arrows	up or down arrow to lock in blue direction; right to lock in red; left to lock in green
	Resize	resize model: measure a distance, type intended size, and Enter
Zoom (Z)	Shift	hold down and click-drag mouse to change Field of View

© 2010 Google Inc.

(Used by permission of Google Inc.)

Appendix B
Ruby Class and Method Reference

The following is a selection of the most commonly used Ruby classes and methods as included with the current version of SketchUp (Version 8, Maintenance Release 3). Some contain SketchUp-specific methods.

```
Array (Parent: Object)
.& .* .+ .- .<< .[] .[]= .all? .any? .assoc .at .clear .collect .collect!
.compact .compact! .concat .cross .delete .delete_at .delete_if .detect
.distance .distance_to_line .distance_to_plane .dot .each .each_index
.each_with_index .empty? .entries .fetch .fill .find .find_all .first
.flatten .flatten! .get_bounds_2d .get_point2d .get_point3d .grep .index
.indexes .indices .inject .insert .join .last .length .map .map! .max
.member? .min .move .nitems .normalize .normalize! .occlusion .offset
.offset! .offsetPoints .on_line? .on_plane? .pack .partition .pop
.project_to_line .project_to_plane .push .rassoc .reject .reject!
.replace .reverse .reverse! .reverse_each .rindex .round_to
.sameConstraint? .select .shift .size .slice .slice! .sort .sort! .sort_by
.to_ary .to_ptr .transform .transform! .transpose .uniq .uniq! .uniq_LDD
.unshift .values_at .vector_to .x .x= .y .y= .z .z= .zip .|
```

```
File (Parent: IO)
.<< .all? .any? .atime .binmode .chmod .chown .close .close_read
.close_write .closed? .collect .ctime .detect .each .each_byte .each_line
.each_with_index .entries .eof .eof? .fcntl .fileno .find .find_all .flock
.flush .fsync .getc .gets .grep .inject .ioctl .isatty .lineno .lineno=
.lstat .map .max .member? .min .mtime .partition .path .pid .pos .pos=
.print .printf .putc .puts .read .read_nonblock .readchar .readline
.readlines .readpartial .reject .reopen .rewind .seek .select .sort
.sort_by .stat .sync .sync= .sysread .sysseek .syswrite .tell .to_i .to_io
.to_ptr .truncate .tty? .ungetc .write .write_nonblock .zip
```

```
Hash (Parent: Object)
.[] .[]= .all? .any? .clear .collect .default .default= .default_proc
.delete .delete_if .detect .each .each_key .each_pair .each_value
.each_with_index .empty? .entries .fetch .find .find_all .grep .has_key?
.has_value? .index .indexes .indices .inject .invert .key? .keys .length
```

```
.map .max .member? .merge .merge! .min .partition .rehash .reject
.reject! .replace .select .shift .size .sort .sort_by .store .to_hash
.update .value? .values .values_at .zip
```

IO (Parent: Object)

```
.<< .all? .any? .binmode .close .close_read .close_write .closed? .collect
.detect .each .each_byte .each_line .each_with_index .entries .eof .eof?
.fcntl .fileno .find .find_all .flush .fsync .getc .gets .grep .inject
.ioctl .isatty .lineno .lineno= .map .max .member? .min .partition .pid
.pos .pos= .print .printf .putc .puts .read .read_nonblock .readchar
.readline .readlines .readpartial .reject .reopen .rewind .seek .select
.sort .sort_by .stat .sync .sync= .sysread .sysseek .syswrite .tell .to_i
.to_io .to_ptr .tty? .ungetc .write .write_nonblock .zip
```

Kernel

```
.Array .Float .Integer .String .` .abort .at_exit .binding .block_given?
.callcc .caller .catch .chomp .chomp! .chop .chop! .eval .exec .exit .exit!
.fail .fork .format .getc .gets .global_variables .gsub .gsub! .iterator?
.lambda .load .local_variables .loop .method_missing .open .p .print
.printf .proc .putc .puts .raise .rand .readline .readlines .require .scan
.select .set_trace_func .sleep .split .sprintf .srand .sub .sub! .syscall
.system .test .throw .trace_var .trap .untrace_var .warn
```

Math

```
.acos .acosh .asin .asinh .atan .atan2 .atanh .cos .cosh .erf .erfc .exp
.frexp .hypot .ldexp .log .log10 .sin .sinh .sqrt .tan .tanh
```

String (Parent: Object)

```
.% .* .+ .<< .[] .[]= .all? .any? .between? .capitalize .capitalize!
.casecmp .center .chomp .chomp! .chop .chop! .collect .concat .count
.crypt .delete .delete! .detect .downcase .downcase! .dump .each
.each_byte .each_line .each_with_index .empty? .entries .find .find_all
.grep .gsub .gsub! .hex .index .inject .insert .intern .length .ljust
.lstrip .lstrip! .map .match .max .member? .min .next .next! .oct
.partition .reject .replace .reverse .reverse! .rindex .rjust .rstrip
.rstrip! .scan .select .size .slice .slice! .sort .sort_by .split .squeeze
.squeeze! .strip .strip! .sub .sub! .succ .succ! .sum .swapcase .swapcase!
.to_f .to_f_LDD .to_i .to_l .to_l_LDD .to_ptr .to_str .to_sym .tr .tr!
.tr_s .tr_s! .unpack .upcase .upcase! .upto .zip
```

Time (Parent: Object)

```
.+ .- ._dump .asctime .between? .ctime .day .dst? .getgm .getlocal .getutc
.gmt? .gmt_offset .gmtime .gmtoff .hour .isdst .localtime .mday .min .mon
.month .sec .strftime .succ .to_f .to_i .tv_sec .tv_usec .usec .utc .utc?
.utc_offset .wday .yday .year .zone
```

Appendix C
SketchUp API Class and Method Reference

This is a complete listing of all classes and methods as they are included in the version of SketchUp that was current as of this writing (Version 8, Maintenance Release 3).

```
Animation (Parent: Object)
.nextFrame .pause .resume .stop
```

```
AppObserver (Parent: Object)
.onNewModel .onOpenModel .onQuit .onUnloadExtension
```

```
ArcCurve (Parent: Curve)
.center .end_angle .normal .plane .radius .start_angle .xaxis .yaxis
```

```
Array (Parent: Object)
.cross .distance .distance_to_line .distance_to_plane .dot .normalize
.normalize! .offset .offset! .on_line? .on_plane? .project_to_line
.project_to_plane .transform .transform! .vector_to .x .x= .y .y= .z .z=
```

```
AttributeDictionaries (Parent: Object)
.[] .delete .each
```

```
AttributeDictionary (Parent: Object)
.[] .[]= .delete_key .each .each_key .each_pair .keys .length .name .size
.values
```

```
Behavior (Parent: Entity)
.always_face_camera= .always_face_camera? .cuts_opening= .cuts_opening?
.is2d= .is2d? .no_scale_mask= .no_scale_mask? .shadows_face_sun=
.shadows_face_sun? .snapto .snapto=
```

```
BoundingBox (Parent: Object)
.add .center .clear .contains? .corner .depth .diagonal .empty? .height
.intersect .max .min .new .valid? .width
```

Camera (Parent: Object)
.aspect_ratio .aspect_ratio= .description .description= .direction .eye
.focal_length .focal_length= .fov .fov= .height .height= .image_width
.image_width= .new .perspective= .perspective? .set .target .up .xaxis
.yaxis .zaxis

Color (Parent: Object)
.alpha .alpha= .blend .blue .blue= .green .green= .names .new .red .red=
.to_a .to_i .to_s

Command (Parent: Object)
.large_icon .large_icon= .menu_text .menu_text= .new .set_validation_proc
.small_icon .small_icon= .status_bar_text .status_bar_text= .tooltip
.tooltip=

ComponentDefinition (Parent: Drawingelement)
.<=> .== .add_observer .behavior .count_instances .description
.description= .entities .group? .guid .hidden? .image? .insertion_point
.insertion_point= .instances .internal? .invalidate_bounds .name .name=
.path .refresh_thumbnail .remove_observer .save_as .save_thumbnail

ComponentInstance (Parent: Drawingelement)
.add_observer .definition .definition= .equals? .explode .glued_to
.glued_to= .intersect .locked? .locked= .make_unique .manifold? .move!
.name .name= .outer_shell .remove_observer .show_differences .split
.subtract .transform! .transformation .transformation= .trim .union
.volume

ConstructionLine (Parent: Drawingelement)
.direction .direction= .end .end= .position .position= .reverse! .start
.start= .stipple .stipple=

ConstructionPoint (Parent: Drawingelement)
.position

Curve (Parent: Entity)
.count_edges .each_edge .edges .first_edge .is_polygon? .last_edge
.length .move_vertices .vertices

DefinitionList (Parent: Entity)
.[] .add .add_observer .at .count .each .length .load .load_from_url
.purge_unused .remove_observer .unique_name

DefinitionObserver (Parent: Object)
.onComponentInstanceAdded .onComponentInstanceRemoved .onComponentAdded

.onComponentPropertiesChanged .onComponentRemoved
.onComponentTypeChanged

Drawingelement (Parent: Entity)
.bounds .casts_shadows= .casts_shadows? .erase! .hidden= .hidden? .layer
.layer= .material .material= .receives_shadows= .receives_shadows?
.visible= .visible?

Edge (Parent: Drawingelement)
.all_connected .common_face .curve .end .explode_curve .faces .find_faces
.length .line .other_vertex .reversed_in? .smooth= .smooth? .soft= .soft?
.split .start .used_by? .vertices

EdgeUse (Parent: Entity)
.edge .end_vertex_normal .face .loop .next .partners .previous .reversed?
.start_vertex_normal

Entities (Parent: Object)
.[] .add_3d_text .add_arc .add_circle .add_cline .add_cpoint .add_curve
.add_edges .add_face .add_faces_from_mesh .add_group .add_image
.add_instance .add_line .add_ngon .add_observer .add_text .at .clear!
.count .each .erase_entities .fill_from_mesh .intersect_with .length
.model .parent .remove_observer .transform_by_vectors .transform_entities

EntitiesObserver (Parent: Object)
.onContentsModified .onElementAdded .onElementModified .onElementRemoved
.onEraseEntities

Entity (Parent: Object)
.add_observer .attribute_dictionaries .attribute_dictionary
.delete_attribute .deleted? .entityID get_attribute .model .parent
.remove_observer .set_attribute .to_s .typename .valid?

EntityObserver (Parent: Object)
.onChangeEntity .onEraseEntity

ExtensionsManager (Parent: Object)
.[] .each .keys .size

Face (Parent: Drawingelement)
.all_connected .area .back_material .back_material= .classify_point
.edges .followme .get_UVHelper .get_glued_instances .loops .material
.material= .mesh .normal .outer_loop .plane .position_material .pushpull
.reverse! .vertices

`Geom` (Parent: N/A)
`.closest_points` `.fit_plane_to_points` `.intersect_line_line`
`.intersect_line_plane` `.intersect_plane_plane` `.linear_combination`
`.point_in_polygon_2D`

`Group` (Parent: Drawingelement)
`.add_observer` `.copy` `.description` `.description=` `.entities` `.equals?` `.explode`
`.intersect` `.local_bounds` `.locked?` `.locked=` `.make_unique` `.manifold?` `.move!`
`.name` `.name=` `.outer_shell` `.remove_observer` `.show_differences` `.split`
`.subtract` `.to_component` `.transform!` `.transformation` `.transformation=`
`.trim` `.union` `.volume`

`Image` (Parent: Drawingelement)
`.explode` `.height` `.height=` `.normal` `.origin` `.origin=` `.path` `.pixelheight`
`.pixelwidth` `.size=` `.transform!` `.width` `.width=` `.zrotation`

`Importer` (Parent: Object)
`.description` `.do_options` `.file_extension` `.id` `.load_file` `.supports_options?`

`InputPoint` (Parent: Object)
`.==` `.clear` `.copy!` `.degrees_of_freedom` `.depth` `.display?` `.draw` `.edge` `.face`
`.new` `.pick` `.position` `.tooltip` `.transformation` `.valid?` `.vertex`

`InstanceObserver` (Parent: Object)
`.onClose` `.onOpen`

`LatLong` (Parent: Object)
`.latitude` `.longitude` `.new` `.to_a` `.to_s` `.to_utm`

`Layer` (Parent: Entity)
`.<=>` `.==` `.name` `.name=` `.page_behavior` `.page_behavior=` `.visible=` `.visible?`

`Layers` (Parent: Entity)
`.[]` `.add` `.add_observer` `.at` `.count` `.each` `.length` `.purge_unused`
`.remove_observer` `.unique_name`

`LayersObserver` (Parent: Object)
`.onCurrentLayerChanged` `.onLayerAdded` `.onLayerRemoved` `.onRemoveAllLayers`

`Length` (Parent: Object)
`.<` `.<=` `.<=>` `.==` `.>` `.>=` `.inspect` `.to_f` `.to_s`

`Loop` (Parent: Entity)
`.convex?` `.edges` `.edgeuses` `.face` `.outer?` `.vertices`

Material (Parent: Entity)
.<=> .== .alpha .alpha= .color .color= .display_name . Type .name .name=
.texture .texture= .use_alpha? .write_thumbnail

Materials (Parent: Entity)
.[] .add .add_observer .at .count .current .current= .each .length
.purge_unused .remove .remove_observer

MaterialsObserver (Parent: Object)
.onMaterialAdd .onMaterialChange .onMaterialRefChange .onMaterialRemove
.onMaterialRemoveAll .onMaterialSetCurrent .onMaterialUndoRedo

Menu (Parent: Object)
.add_item .add_separator .add_submenu .set_validation_proc

Model (Parent: Object)
.abort_operation .active_entities .active_layer .active_layer=
.active_path .active_view .add_note .add_observer .attribute_dictionaries
.attribute_dictionary .behavior .bounds .close_active .commit_operation
.definitions .description .description= .edit_transform .entities .export
.georeferenced? .get_attribute .get_datum .get_product_family .guid
.import .latlong_to_point .layers .list_datums .materials .mipmapping=
.mipmapping? .modified? .name .name= .number_faces .options .pages .path
.place_component .point_to_latlong .point_to_utm .raytest
.remove_observer .rendering_options .save .save_thumbnail .select_tool
.selection .set_attribute .set_datum .shadow_info .start_operation
.styles .tags .tags= .title .tools .utm_to_point .valid?

ModelObserver (Parent: Object)
.onActivePathChanged .onAfterComponentSaveAs .onBeforeComponentSaveAs
.onDeleteModel .onEraseAll .onExplode .onPreSaveModel .onPostSaveModel
.onPlaceComponent .onSaveModel .onTransactionAbort .onTransactionCommit
.onTransactionEmpty .onTransactionRedo .onTransactionStart
.onTransactionUndo

Numeric (Parent: Object)
.cm .degrees .feet .inch .km .m .mile .mm .radians .to_cm .to_feet
.to_inch .to_km .to_l .to_m .to_mile .to_mm .to_yard .yard

OptionsManager (Parent: Object)
.[] .count .each .keys .size

OptionsProvider (Parent: Object)
.[] .[]= .add_observer .count .each .each_key .each_pair .each_value
.has_key? .key? .keys .name .remove_observer .size

OptionsProviderObserver (Parent: Object)
.onOptionsProviderChanged

Page (Parent: Entity)
.camera .delay_time .delay_time= .description .description=
.hidden_entities .label .layers .name .name= .rendering_options
.set_visibility .shadow_info .style .transition_time .transition_time=
.update .use_axes= .use_axes? .use_camera= .use_camera? .use_hidden=
.use_hidden? .use_hidden_layers= .use_hidden_layers?
.use_rendering_options= .use_rendering_options? .use_section_planes=
.use_section_planes? .use_shadow_info= .use_shadow_info? .use_style=
.use_style?

Pages (Parent: Entity)
.[] .add .add_frame_change_observer .add_matchphoto_page .add_observer
.count .each .erase .parent .remove_frame_change_observer
.remove_observer .selected_page .selected_page= .show_frame_at .size
.slideshow_time

PagesObserver (Parent: EntitiesObserver)
.onContentsModified .onElementAdded .onElementRemoved

PickHelper (Parent: Object)
.all_picked .best_picked .count .depth_at .do_pick .element_at .init
.leaf_at .path_at .pick_segment .picked_edge .picked_element .picked_face
.test_point .transformation_at .view

Point3d (Parent: Object)
.+ .- .< .== .[] .[]= .clone .distance .distance_to_line
.distance_to_plane .inspect .linear_combination .new .offset .offset!
.on_line? .on_plane? .project_to_line .project_to_plane .set! .to_a .to_s
.transform .transform! .vector_to .x .x= .y .y= .z .z=

PolygonMesh (Parent: Object)
.add_point .add_polygon .count_points .count_polygons .new .normal_at
.point_at .point_index .points .polygon_at .polygon_points_at .polygons
.set_point .transform! .uv_at .uvs

RenderingOptions (Parent: Object)
.[] .[]= .add_observer .each .each_key .each_pair .keys .remove_observer

RenderingOptionsObserver (Parent: Object)
.onRenderingOptionsChanged

SectionPlane (Parent: DrawingElement)
.get_plane .set_plane

Selection (Parent: Object)
.[] .add .add_observer .at .clear .contains? .count .each .empty? .first
.include? .is_curve? .is_surface? .length .model .nitems .remove
.remove_observer .shift .single_object? .toggle

SelectionObserver (Parent: Object)
.onSelectionAdded .onSelectionBulkChange .onSelectionCleared
.onSelectionRemoved

Set (Parent: Object)
.clear .contains? .delete .each .empty? .include? .insert .length .new
.size .to_a

ShadowInfo (Parent: Entity)
.[] .[]= .add_observer .each .each_key .each_pair .keys .remove_observer

ShadowInfoObserver (Parent: Object)
.onShadowInfoChanged

Sketchup (Parent: N/A)
.active_model .add_observer .app_name .break_edges= .break_edges?
.create_texture_writer .display_name_from_action .extensions .file_new
.find_support_file .find_support_files .format_angle .format_area
.format_degrees .format_length .full_detail_render_delay_max=
.full_detail_render_delay_min= .fix_shadow_strings= .fix_shadow_strings?
.get_datfile_info .get_i18ndatfile_info .get_locale .get_resource_path
.get_shortcuts .install_from_archive .is_online .is_pro?
.is_valid_filename? .load .open_file .os_language .parse_length
.plugins_disabled= .plugins_disabled? .read_default .register_extension
.register_importer .remove_observer .require .save_thumbnail .send_action
.set_status_text .status_text= .template .template= .template_dir .und
.vcb_label= .vcb_value= .version .version_number .write_default

SketchupExtension (Parent: Object)
.check .copyright .copyright= .creator .creator= .description
.description= .load_on_start? .loaded? .name .name= .new .registered?
.uncheck .version .version=

String (Parent: Object)
.to_l

Style (Parent: Object)
.description .description= .name .name=

Styles (Parent: Object)
.[] .active_style .active_style_changed .add_style .count .each .parent
.purge_unused .selected_style .selected_style= .size
.update_selected_style

Text (Parent: DrawingElement)
.arrow_type .arrow_type= .display_leader= .display_leader? .has_leader?
.leader_type .leader_type= .line_weight .line_weight= .point .point=
.set_text .text .text= .vector .vector=

Texture (Parent: Entity)
.average_color .filename .height .image_height .image_width .size= .valid?
.width

TextureWriter (Parent: Object)
.count .filename .handle .length .load .write .write_all

Tool (Parent: Object)
.activate .deactivate .draw .enableVCB? .getExtents
.getInstructorContentDirectory .getMenu .onCancel .onKeyDown .onKeyUp
.onLButtonDoubleClick .onLButtonDown .onLButtonUp .onMButtonDoubleClick
.onMButtonDown .onMButtonUp .onMouseEnter .onMouseLeave .onMouseMove
.onRButtonDoubleClick .onRButtonDown .onRButtonUp .onReturn .onSetCursor
.onUserText .resume .suspend

Toolbar (Parent: Object)
.add_item .add_separator .each .get_last_state .hide . name .new .restore
.show .visible?

Tools (Parent: Object)
.active_tool_id .active_tool_name .add_observer .model .pop_tool
.push_tool .remove_observer

ToolsObserver (Parent: Object)
.onActiveToolChanged .onToolStateChanged

Transformation (Parent: Object)
.* .axes .clone .identity? .interpolate .inverse .invert! .new .origin
.rotation .scaling .set! .to_a .translation .xaxis .yaxis .zaxis

UI (Parent: N/A)
.add_context_menu_handler .beep .create_cursor .inputbox
.inspector_names .menu .messagebox .model_info_pages .openURL .openpanel
.play_sound .preferences_pages .refresh_inspectors .savepanel .set_cursor
.set_toolbar_visible .show_inspector .show_model_info .show_preferences
.start_timer .stop_timer .toolbar .toolbar_names .toolbar_visible?

UVHelper (Parent: Object)
.get_back_UVQ .get_front_UVQ

Vector3d (Parent: Object)
.% .* .+ .- .< .== .[] .[]= .angle_between .axes .clone .cross .dot
.inspect .length .length= .linear_combination .new .normalize .normalize!
.parallel? .perpendicular? .reverse .reverse! .samedirection? .set! .to_a
.to_s .transform .transform! .unitvector? .valid? .x .x= .y .y= .z .z=

Vertex (Parent: Entity)
.common_edge .curve_interior? .edges .faces .loops .position .used_by?

View (Parent: Object)
.add_observer .animation= .average_refresh_time .camera .camera= .center
.corner .draw .draw2d .draw_line .draw_lines .draw_points .draw_polyline
.draw_text .drawing_color= .dynamic= .field_of_view .field_of_view=
.force_invalidate .guess_target .inference_locked? .inputpoint .invalidate
.last_refresh_time .line_stipple= .line_width= .lock_inference .model
.pick_helper .pickray .pixels_to_model .refresh .remove_observer
.screen_coords .set_color_from_line .show_frame .tooltip= .vpheight
.vpwidth .write_image .zoom .zoom_extents

ViewObserver (Parent: Object)
.onViewChanged

WebDialog (Parent: Object)
.add_action_callback .allow_actions_from_host .bring_to_front .close
.execute_script .get_default_dialog_color .get_element_value .max_height
.max_height= .max_width .max_width= .min_height .min_height= .min_width
.min_width= .navigation_buttons_enabled= .navigation_buttons_enabled?
.new .post_url .set_background_color .set_file .set_full_security=
.set_html .set_on_close .set_position .set_size .set_url .show
.show_modal .visible? .write_image

Appendix D
Creating Your Own Plugins

The following is a basic template to get you started with writing your own plugins for SketchUp. Even if you don't want to publish your plugin, it might be useful sometimes to "install" a script you created in SketchUp using this convenient method.

RBZ Plugin File Structure

The RBZ file format was introduced in Maintenance Release 2 of Version 8. Packaging a plugin this way allows users to easily install it from SketchUp's Preferences dialog.

To enable use of the RBZ-installer functionality, all files that make up a plugin must be packaged in one single compressed file (note that the RBZ file format is simply a renamed ZIP file). The following is a minimal sample file structure. You can add as many other files and folders as necessary for your plugin. The entire contents of the compressed file (including the folder structure) is copied into SketchUp's Plugins folder on installation and the plugin is made available immediately.

```
my_plugin.rbz (compressed file)
  |
  |— my_plugin_loader.rb
  |
  |— my_plugin (directory)
      |
      |— my_plugin.rb
```

> **TIP**
>
> As a reference: SketchUp's default plugin installation folder can be found here (replace version number with current major version):
>
> Windows: C:\Program Files\Google\Google SketchUp 8\Plugins\
>
> Mac: /Library/Application Support/Google SketchUp 8/SketchUp/Plugins/

Plugin Template Structure

It is a good idea to (1) use SketchUp's extension system (so that plugins can be turned on and off by the user) and (2) wrap your plugin in its own Ruby Module (to prevent conflicts between installed plugins). Following is some sample code to get you started. Replace "my" with your initials to identify you as the author and keep things clean.

Contents of **/my_plugin_loader.rb** (in main directory):

```
require "sketchup.rb"
require "extensions.rb"

# Load plugin as extension (so that user can disable it)

my_plugin_loader = SketchupExtension.new "My_Plugin Loader",
"my_plugin/my_plugin.rb"
my_plugin_loader.copyright= "Copyright 2012 by Me"
my_plugin_loader.creator= "Me, myself and I"
my_plugin_loader.version = "1.0"
my_plugin_loader.description = "Description of plugin."
Sketchup.register_extension my_plugin_loader, true
```

Contents of **/my_plugin/my_plugin.rb** (in subfolder):

```
=begin
Copyright 2012, Author
All Rights Reserved
THIS SOFTWARE IS PROVIDED "AS IS" AND WITHOUT ANY EXPRESS OR IMPLIED
WARRANTIES, INCLUDING, WITHOUT LIMITATION, THE IMPLIED WARRANTIES OF
MERCHANTABILITY AND FITNESS FOR A PARTICULAR PURPOSE.
License: AuthorsLicenseStatement
Author: AuthorName
Organization: AuthorAffiliationOrOrganizationIfAny
Name: ScriptName
Version: ScriptVersion
SU Version: MinimumSketchUpVersion
Date: Date
Description: ScriptDescription
Usage: ScriptUsageInstructions
History:
 1.000 YYYY-MM-DD Description of changes
=end

require "sketchup.rb"
# Main code (start module name with capital letter)
module My_module
 def self.my_method
 # do something...
 end
 def self.my_second_method
```

```
  # do something...
  end
end
# Create menu items
unless file_loaded?(__FILE__)
  mymenu = UI.menu("Plugins").add_submenu("My Plugin Collection")
  mymenu.add_item("My Tool 1") {My_module::my_method}
  mymenu.add_item("My Tool 2") {My_module::my_second_method}
  file_loaded(__FILE__)
end
```

TIP

If you prefer not to have your plugin's source code visible, then you can use SketchUp's Scrambler to encrypt Ruby files. You can download it from here:

https://developers.google.com/sketchup/docs/utilities

Appendix E
Dynamic Component Function Reference

The following list includes all of the Dynamic Component functions available in the current SketchUp release (Version 8, Maintenance Release 3):

Predefined Attributes:

```
X, Y, Z, LenX, LenY, LenZ, RotX, RotY, RotZ, Hidden, Copies, Copy, Name,
Summary, Description, Creator, ItemCode, ImageURL, DialogWidth,
DialogHeight, onClick, Material
```

Math Functions:

```
ABS(number), CEILING(number, significance), DEGREES(number), EVEN(number),
EXP(number), FLOOR(number, significance), INT(number), ISEVEN(number),
ISODD(number), LN(number), LOG10(number), ODD(number), PI(),
RADIANS(number), RAND(), RANDBETWEEN(bottom, top), ROUND(number, count),
SIGN(number), SQRT(number)
```

SketchUp Functions:

```
CHOOSE(index,value1,value2, ...valueN), CURRENT("attributeName"), EDGES(),
FACEAREA("materialName"), FACES(), LARGEST(value1,value2,...valueN), LAT(),
LNG(), NEAREST(originalValue, value1, value2, ...valueN),
OPTIONINDEX("attributeName"), OPTIONLABEL("attributeName"),
SMALLEST(value1,value2,...valueN), SUNANGLE(), SUNELEVATION()
```

Text Functions:

```
CHAR(number), CODE(text), CONCATENATE(text1, text2, ...textN),
DOLLAR(value, decimals), EXACT(text1, text2), FIND(findText, text,
position), LEFT(text, number), LEN(text), LOWER(text), MID(text, start,
number), PROPER(text), REPLACE(text, position, length, new), REPT(text,
number), RIGHT(text, number), SUBSTITUTE(text, searchText, newText,
occurrence), TRIM(text), UPPER(text), VALUE(text)
```

Trigonometric Functions:

```
ACOS(number), ACOSH(number), ASIN(number), ASINH(number), ATAN(number),
ATANH(number), COS(number), COSH(number), SIN(number), SINH(number),
TAN(number), TANH(number)
```

Logical Functions:

```
AND(logicalValue1, logicalValue2, ...logicalValueN), FALSE(), IF(test,
thenValue, elseValue), NOT(logicalValue), OR(logicalValue1, logicalValue2,
...logicalValueN), TRUE()
```

OnClick Functions:

```
ALERT("message"), ANIMATE(attribute, state1, state2, ... stateN),
ANIMATESLOW(attribute, state1, state2, ... stateN), ANIMATEFAST(attribute,
state1, state2, ... stateN), ANIMATECUSTOM("attribute", time, easein,
easeout, state1, ...stateN), GOTOSCENE("sceneName," time, easein, easeout),
REDRAW(), SET("attribute", state1, state2, ...stateN)
```

Supported Operators:

```
+ (add), - (subtract), * (multiply), / (divide), < (less than), > (greater
than), <= (less than or equal to), >= (greater than or equal to), =
(equal), () (parentheses), <> (not equal to)
```

Index